THE GOLD OF OPHIR

FACSIMILE OF A HIMYARITIC ROCK INSCRIPTION : GLASER NO. 282.

(To be read from right to left, see p. 164.)

THE GOLD OF OPHIR

Whence brought and by Whom?

BY

PROFESSOR A. H. KEANE, F.R.G.S.

'The truth of Ophir must, as from a deep myne, be drawne out of Moses, Gen. 10'—PURCHAS, I. 26

'There was a true Ophir, named of Ophir, the son of Joktan'—IB. *p.* 45

'Seetzen has conclusively shown that Ophir is to be looked for in Southern Arabia'

BURTON: *The Land of Midian,* II. *p.* 239

NEGRO UNIVERSITIES PRESS
NEW YORK

Originally published in 1901
by Edward Stanford, London

Reprinted 1969 by
Negro Universities Press
A Division of Greenwood Publishing Corp.
New York

SBN 8371-1725-9

PRINTED IN UNITED STATES OF AMERICA

PREFACE

A FEW words seem called for to explain how I came to risk my reputation for sanity by plunging into this 'Ophir Question,' so often authoritatively declared to be insoluble. When such pronouncements were made it was insoluble, because some of the essential factors of the problem were missing. Such is no longer the case, and during the last three decades, that is, since the re-discovery of the Zimbabye monuments in the present Rhodesia, materials have been accumulated from various quarters, which justify the re-opening of the subject, and have seemed to me amply sufficient for its final settlement. These fresh materials fall under several heads, the most important of which may here be specified.

First and foremost come the extensive though still far from exhausted explorations and careful studies of the Rhodesian remains, together with general surveys of the whole ground, by thoroughly competent observers—trained archæologists such as the late Theodore Bent and Mr. Robert Swan; practical miners, engineers, and men of high scientific knowledge, such as Dr. Henry Schlichter, Dr. Carl Mauch, Mr. Thomas Baines, Mr. E. A. Maund and Mr. Franklin White; lastly experienced local explorers, such as Messrs. Hall and Neal with their worthy associate, Mr. Johnson, who have spent years of intelligent labour investigating the whole of this wonderful auriferous region, mapping its hundreds of ancient gold workings and classifying the associated 'Zimbabyes' into periods and types. Thus was given the indispensable clue to the time sequences, and to the architectural prototypes if any could elsewhere be found, for all these observers have from the first been of accord that none of the early and more finished structures, those especially of the first and second periods, could be ascribed to the present Bantu populations of Rhodesia. And such an attribution is now altogether excluded, since I have been able to show that these Negroid Bantu peoples were preceded by a still lower race—the Bushman-Hottentots—who occupied the land at the very time the monuments were raised, and indeed supplied the forced labour necessary for their profitable erection.

While these operations were in progress, or even prior to them, others, notably Joseph Halévy, Lieutenant J. R. Wellsted, Eduard Glaser, Julius Euting, Thomas Arnaud, Siegfried Lander, and the pioneer, Christian

Seetzen, mostly eminent orientalists and accomplished archæologists, were at work amid the ruins and inscribed rocks thickly strewn over Southern Arabia. Nearly two thousand inscriptions in two forms of the ancient Himyaritic language—Minaean and Sabaean—were recovered and partly deciphered, while numerous groups of crumbling monuments were surveyed and described. These descriptions arrested immediate attention, and by the comparative method of study the monuments themselves were soon brought into line with those beyond the Zambesi. Further research multiplied the points of contact both in the forms and material of the structures, and in the objects brought to light from their débris. The parallelisms ceased to be coincidences and became identities; the elliptical temples at the ancient Sabaean capital, Maraiaba, at Nakb el-Hajar and elsewhere, presented the most striking resemblance to several of the Zimbabyes; the symbols figured on a Phœnician coin of Byblos looked like miniature plans of certain South African groups; an ingot of tin found in Falmouth Harbour might have been cast from a soapstone mould of curious form brought from Great Zimbabye; the very date of this structure (1100 B.C.) has been astronomically determined by a zodiacal chart which was found in the neighbourhood and proved to be the work of northern star-gazers dating from a time when Sol entered Taurus at the vernal equinox. All this cumulative evidence left no doubt that the foreign prototypes of the Rhodesian monuments had been found in the Himyaritic lands of Southern Arabia, seat, as some now think, of the oldest civilisation in the world. Else we are entitled to ask, If the South African buildings, all intimately connected with gold-winning, were not raised by the Himyarites and their Phœnician kinsmen, by whom were they raised? Did they drop from the clouds? And even so, how came they to simulate the architectural style of their Yemenite counterparts? Miracle upon miracle has to be suggested by the sceptics, if any remain, to avoid a very obvious and inevitable inference.

And again, why simulate on the stones and earthenware of Zimbabye the very script recurring on the rock inscriptions of Arabia Felix? The texts of these graven stones enter as a third factor of vital importance in the problem. They reveal long lists of Sabaean kings dating at least from the time of Solomon and Hiram, and other very much older lists of Minaean kings—thirty-three have so far been recovered—going back to a dim past coeval with the early Egyptian and Babylonian epochs. They further show that the art of navigation was already well developed in those remote times, that the Babylonians probably acquired the art from the Himyarites already dominant in the Indian Ocean; and that there were Minaean potentates who offered their supplications to the land gods and the sea gods of all the regions with which they had established intercourse.

A fourth factor was yielded unexpectedly, and indeed unwittingly, by Bent's explorations in the South Arabian frankincense land, Dhofar. Here are concentrated in strange profusion all the elements needed to establish the identity of Moscha—Arrian's Portus Nobilis—and the ruins grouped round the adjacent inlet, with the Biblical Ophir, Ptolemy's Sapphar Metropolis under the very shadow of the Mount Sephar of Genesis x. The survey of the district showed that the famous harbour, running nearly two miles inland, has long been silted up at its mouth, thus explaining how it came to be forgotten throughout mediæval times till now again recovered by modern enterprise. With its recovery Ophir is also recovered, and is found to be not a gold-yielding land, but a gold mart, a gold importer and distributor throughout the ancient world, as might indeed have been anticipated by a more careful study of the Biblical texts. But as long as the idea persisted that Ophir was itself an auriferous land it would never have been discovered, as I trust has now been made clear. It is here also made clear (see Chapter IX.) that the Biblical Tharshish, the Tharshish of Solomon and Hiram, is to be sought in the Indian Ocean, and not at the other end of the Mediterranean, as might be inferred from the vagaries of the Septuagint and Vulgate (Greek and Latin) translations of certain passages in Kings and Chronicles (see Note 38). How any intelligent commentator could ever have supposed that Solomon's Tharshish was the Tartessus at the mouth of the Guadalquivir on the Atlantic seaboard, passes comprehension. Why should the Israelitish and Phœnician navies, built at Solomon's naval station of Ezion-geber on the Red Sea, try to get through the Isthmus of Suez to sail down the whole length of the Mediterranean in quest of the ivory and peacocks which they never could find in Spain but could easily ship at various ports of the Indian Ocean? If Tartessus was their objective, they could have sailed straight from Tyre, and Solomon need not have troubled about Ezion-geber, which nevertheless plays such a conspicuous part in the naval records of his reign.

A fifth factor of great interest and novelty is introduced for the first time into the question in Chapters XII. and XIII., where long-standing social and religious intercourse is established between the Malagasy inhabitants of Madagascar and their Himyaritic, Phœnician, and Jewish visitors from the northern hemisphere. To reach these southern latitudes, almost within sight of Rhodesia, the Himyarites had only to start from their port of Ophir, whence they were already accustomed to send expeditions to all parts of the Indian Ocean. But their Phœnician and Israelitish associates had to sail down the whole length of the Red Sea, and then round by the south coast of Arabia to Ophir, which lay about midway between Ezion-geber and the Mozambique Channel. It may be gathered from the Biblical texts that many of the expeditions stopped here, while others, especially those equipped by Solomon and Hiram jointly, passed on to Tharshish, port of the auriferous Havilah (Rhodesia). In this way all these Semitic peoples—Himyarites, Jews, and Phœnicians—may well have found their way to the great island, where their long sojourn is still attested by the numerous Himyaritic elements in the Malayo-Polynesian Malagasy language, especially terms connected with the Sabaeo-Babylonian astrology, and divination, as well as by the survival of many Canaanitish religious observances and Levitical rites prevalent in Phœnicia and Israel in Solomonic times. It will come as a surprise to many of my readers to learn that the Malagasy people name their week-days, not in the relatively modern Arabic of the Koran, but in the far more ancient Himyaritic language of the rock inscriptions, and that all their month-names are those not of the Moslem Arabs, but of the Sabaeo-Babylonian Zodiacal constellations, and that not in translations, as we might say the Ram, the Bull, the Twins, &c., but in the very Semitic forms themselves, slightly modified in conformity with the Malagasy phonetic system. It will also be a surprise to folklorists to find that many of the popular myths, legends, nursery rhymes and stories, are strikingly analogous to those current from time out of mind amongst the European populations, and that the explanation lies probably in the gold-hunting expeditions despatched by Solomon and his friend Hiram to the southern waters some three thousand years ago. But such popular fancies seem never to die, but still to gain new life and put on endless Protean forms with every rebound from mother earth. Nor will anybody feel much surprise at the vitality of this remarkable Malagasy folklore, when he has studied the proofs here submitted of the still more astonishing vitality of the Malagasy language itself (see p. 132 *sq.* and Note 66).

It must now be pretty evident to everybody that the Ophir problem was necessarily insoluble before this great mass of fresh evidence had been placed at the disposal of the historical student.

The field here surveyed is so extensive, and the materials now available so abundant, that in order to prevent overcrowding I have had to relegate much important matter to the 'Notes and Addenda,' collected together at the end of the volume. These are therefore to be taken as an integral part of the argument, and many may be described as what the French call *pièces justificatives.* I therefore trust that my critics will carefully consult them before venturing an opinion on the merits of any point at issue.

During the progress of the work I have been in correspondence with Messrs. Hall and Neal, and must here tender them my warmest thanks for their courtesy and generosity in placing at my disposal some of the advance sheets of the 'Ancient Ruins of Rhodesia,' in which they have embodied the results of the important investigations above referred to. In Chapters II. and XIV. I have gladly availed myself of some of the valuable materials collected by them. Needless to say that their work will be found indispensable for a full understanding of the whole subject.

I should add that this volume has grown out of a series of papers on 'The Gold of Ophir' communicated to 'Rhodesia' during the months of August and September 1901.

A. H. KEANE.

ARÁM-GÁH:
79 BROADHURST GARDENS, N.W.
December, 1901.

CONTENTS

THE GOLD OF OPHIR

CHAPTER I

DE BARROS' 'KINGDOM OF SOFALA'—THE MONOMOTAPA AND HIS ZIMBABYES

WHEN the Portuguese first rounded the Cape, and swept tumultuously into the eastern seas, they hurried, like their Spanish rivals in the Western Hemisphere, everywhere in quest of gold and the other treasures of the Orient. On capturing Sofala, in 1505, they at once built a stronghold which they called Fort Ophir, because they found the 'Moors,' that is, the local Arab traders, trafficking in gold, which they immediately concluded must be the 'Gold of Ophir.' In this they were mainly right, though wrong, as will be seen, in supposing that the neighbouring auriferous region—the present Rhodesia—was also the land of Ophir. As already suggested by our title, the gold did not come primarily from Ophir, but was brought thither from over the seas; and our present purpose will be to show that the first purveyors of the precious metals, and of the other costly wares brought to the courts of David and Solomon, were the Himyaritic people of South Arabia, whose chief emporium was Ophir, and whose chief foreign settlement was the 'Ancient Rhodesia,' here identified with the Havilah of Scripture—'the whole land of Havilah, where there is gold.' Hence for the '*Sofala*, thought Ophir,' of Milton (P. L. xi), we have merely to substitute *Havilah* to grasp the true relations.

The Portuguese pioneers did not, with some distinguished modern philologists, connect Sofala itself with Ophir, as thus: *Sofala*, *Sofal*, *Sofar*, *Ofar*, *Ophir*; but they thought their general conclusion greatly strengthened when they heard that the largest of the neighbouring coast streams was called *Sabi*, as it still is. They almost inevitably associated this name with that of the Queen of Saba (Sheba), the more so since they were further informed that Sofala had for ages been the great outlet for the gold brought down from the old mines of Manica and the other auriferous inland districts. At that time most of the gold-fields appear to have been comprised within the limits of a single state, the 'Kingdom of Sofala,' as it was called, which was ruled over by a powerful chief named Benamatápa (Benomotápa), or Monomotápa, who is already mentioned by Duarte Barbosa (1512), and again by Camoens in x. 93:

Vê do Benomotápa o grande imperio
De salvatica gente, negra e nua.'

Much valuable information regarding this potentate, his gold mines, and now ruined royal residences (*Sym-*

baoe, Zimbaoé, Zimbabwe, Zimbabye), has been preserved by the contemporary historian De Barros (1496–1570) in his 'Asia,' First Decade, Book X. chap. i. But the Portuguese text not being accessible to the ordinary reader, I subjoin an English version which I have prepared of those passages which bear more directly on the present subject:

'All the land which we include in the kingdom of Sofala is a great region ruled by a heathen prince called Benomotapa; it is enclosed like an island by two arms of a river which issues from the most considerable lake in all Africa, which was much sought after by the ancients as being the hidden source of the famous Nile, whence also issues our Zaire [Congo] which flows by the kingdom of Congo. . . .

'The river, which flows towards Sofala after issuing from this lake, and has a long course, divides into two branches. One reaches the coast this side Cape Corrientes, and is the river which we formerly called the Rio da Lagoa, and now the Rio do Espirito Santo [Limpopo], lately fixed by Lourenço Marques, who discovered it in 1545, and the other branch debouching twenty-five leagues below [above] Sofala is called the Cuama, although other people inland call it Zambere [Zambese].[1] This branch is much the more copious, being navigable for over 250 leagues, and into it fall these six large rivers —Panhames [Hanyani ?], Luamguoa [Oangwa], Arruya [Ruia], Manjovo [Majova], Inadire [?] and Ruenia [Ruenya], all of which water Benomotapa's land, and most of them carry down much gold which is yielded

[1] See end of Volume for Notes.

by that land. Thus these two branches, with the sea on the other side, form this great kingdom of Sofala into an island, which may have a circuit of over 750 leagues. . . .

'The mines of that region, where gold is obtained, which lie nearest to Sofala, are those called Manica; they are in a district encircled by mountains which may have a circuit of thirty leagues; and the locality which yields the gold is generally known by its dry and arid appearance, and the whole territory is called Matuca [the present Mashona], and the people who mine it are the Botongas.[2]

'In those Manica mines, which may lie some fifty leagues to the west of Sofala,[3] the Caffres [4] have to labour hard, the land being so dry; for all the gold found there is in dust, so that they have to take the excavated earth to a place where water can be had, for which they make some holes where it collects in winter, and generally nobody digs more than six or seven spans deep [four to five feet], and if they go to twenty they come upon the hard rock.

'The other mines farther removed from Sofala may be from 100 to 200 leagues distant; they are in the district of Boro and Quiticuy [Quiteve or Kiteve], and in them, as well as in the above-mentioned rivers which water that land, the gold is found of larger size, some embedded in the reefs, some already cleared by the winter torrents; hence, in some of the pools, such as remain in summer, they dive down and find much gold in the mud brought up. In other places, where are some lagoons, 200 men set to work to drain off about half the water, and in the mud which they sift they also find gold; and so rich is the ground that if the people were industrious great quantities could be had; but they are so indolent that stress of hunger alone will keep them at the mines. Hence the Moors who visit those districts have recourse to a ruse to make them diligent. They deck the negroes and their women with clothes, beads, and trinkets, in which they delight, and when all are pleased, trust everything to them, bidding them to go and work the mines, and on their return at such a time they can pay for those things; so that in this way, by giving them credit, the Moors induce them to work, and so truthful are the negroes that they keep their word.'

This passage is of great importance, as showing that 400 years ago the gold-fields were entirely in the hands of the natives, and only indirectly exploited by the Arab dealers, who visited the district for the purpose. It further appears that the operations were limited to washings in the auriferous alluvial mud, so that the old workings, with their shafts, adits, and other abundant evidences of skilled processes, must be nearly all anterior to that date, though often credited to the Moslem Arab traders. Some few, especially in the eastern (Manica) district, might no doubt be attributed to the Portuguese of a later period. But the workings are far too extensive to be thus explained, and, moreover, occur for the most part in districts such as Matabililand and North Transvaal, which were never reached, or, at least, never permanently occupied, by the Portuguese.

De Barros proceeds: 'There are other mines in a

district called Toroa,[5] which is otherwise known as the kingdom of Butua,[5] whose ruler is a vassal of Benomotapa. This land is near the other which we said consisted of extensive plains [plateaux], and those mines are the oldest known in that region [Rhodesia]. They are all in a plain, in the middle of which stands a square [should be *round*] fortress, all of dressed stones within and without, well wrought and of marvellous [read *small*] size, without any mortar showing the joinings, the walls of which are over 25 hands thick [18 to 19 feet] ; but the height is not so great compared to the thickness. And above the gateway of that edifice is an inscription which some learned Moorish [Arab] traders who had been there were unable to read or say what writing it was. [We shall see that the script was almost certainly Himyaritic, therefore unknown to the Moslem Arabs.]

'All these structures the indigenes call SYMBAOE, which, with them, means a Court [royal residence], for every place where Benomotapa stays is so called ; and as they speak of this as being a royal building, all the other dwellings of the king bear the same name. There is a nobleman in charge of them like an alcalde mór [head constable], and such an officer they call Symbacáyo, as we might say warden of the Symbaoe ; and in it are always some of Benomotapa's women, of whom this Symbacáyo has charge.

'When or by whom these edifices were built, the people of the land, being unlettered, have no memory ; only they say they are the work of the devil, because, compared with their own cunning and skill, they think not men could have made them ; and some Moors who had seen them, when shown the work of our stronghold [Fort Ophir] by Vincente Pegado, former Commander of that place [Sofala]—the workmanship of the windows and arches in comparison with the dressed stonework of those buildings—said they were not to be compared. They lie west of Sofala in a straight line 170 leagues more or less, under the latitude of 20° and 21° south, and there are no other old or modern buildings in those parts, for the people are very barbarous, and all their houses are of wood. In the opinion of the Moors who saw them they seemed to be very ancient, and were built there to hold possession of those mines which are very old, from which for years no gold has been taken owing to the wars. And considering the position and style of the buildings erected so far in the heart of the country, the Moors also admitting that these are not their work, seeing their great age, and still more because they had no knowledge of the letters of the inscription over the gateway, we may well conjecture that to be the region by Ptolemy called Agysymba; for its name somewhat agrees, being partly corrupted one from the other.[6] And grounding our judgment on this, the work seems to have been ordered by some prince who at that time was lord of those mines, in order to hold them, but lost them with time, and also because of their remoteness from his dominions; for, as regards their appearance, they seem much like others in the land of Prester John [Abyssinia] in a place called Acaxumo, which city was a residence of Queen Saba, which Ptolemy calls Axuma [Axum], and the lord of that state was also lord of the mines, and on their account had those structures raised, just as we now hold the fortress of Mina [in Manicaland] and this one of Sofala itself.'

That this description refers to the Rhodesia monuments is self-evident, for they are actually named, and the position 'Great Zimbabye,' near the present Victoria in Matabililand, is correctly given as between 20° and 21° south, due west of Sofala, though the distance (170 leagues) is too great, as is so often the case with the early writers. The object of the buildings—'to hold the mines'— and their workmanship—dressed stones without mortar—are also for the most part correctly stated, while their great age is emphasised on the ground that none of the learned Moors could read the unhappily now vanished inscription over the gateway. Now, all 'learned' Moors could read not only the ordinary Arabic script of the Koran, but also the much older monumental Kufic script from which that cursive style was derived about or soon after the time of Mohammed. Hence the inscription was in a pre-Kufic character, and must consequently be dated back to pre-Mohammedan times, that is, as will be seen, to the period of the very ancient Himyaritic rock inscriptions of South Arabia.

De Barros' reference to the Axumite people of Abyssinia as the possible builders and workers of the mines is most remarkable, and, as will appear, comes very near the truth. The Axumite empire is known to have been founded by the South Arabian Himyarites, who crossed the Red Sea from Arabia Felix (Yemen) at some very remote period, and have to this day been the dominant people in Abyssinia. The present Emperor Menelik himself claims descent from Solomon and the Queen of Sheba, who is here mentioned by De Barros as residing in 'Acaxumo,' that is, Axum, capital of the empire. This is, of course, merely a national Biblical tradition, which dates only from the fourth century of the new era, when the Abyssinian Himyarites were converted to Christianity by their apostle, Frumentius of Alexandria. But if the Rhodesian monuments cannot be credited to these Abyssinian Himyarites, it will here, I trust, be made evident that the builders were their ancestors, the Sabaean and Minaean Himyarites of South Arabia.

De Barros continues: 'The prince whom we call Benomotapa or Monomotapa, is, as with us, an emperor, for this is the meaning of his name amongst them, and his state does not consist of much pomp or show about his person. For insignia of his royal state he has a very small hoe with ivory head, which he always wears in his girdle in sign of peace, and another emblem is one or two assegais, to denote justice and defence of his people.'

We thus see, as I have elsewhere pointed out, that Monomotapa was not a principality but a 'prince,' not an empire but an 'emperor,' and that his kingdom of Sofala formed a sort of island enclosed by the branches of the above-described river. Thus vanishes the fabulous 'empire of Monomotapa,' which was supposed to occupy a great part of the interior, and has in vain been sought by geographers for over three hundred years. Both forms of the word have the same meaning, the first components, *Bena* and *mono*, being the ordinary Bantu words (*bwana*, *bana*, *muene*, *mwana*) for lord, chief, master. The second part evidently means 'mine,' from *tapa*, to dig, excavate ; so that Benomotápa and Monomotápa would mean 'Lord of the Mines,' an appropriate title for the ruler of the auriferous Rhodesian lands.

II

CHAPTER II

BENT'S 'ANCIENT RHODESIA'—ITS MONUMENTS AND GOLD WORKINGS

BUT with the decadence of Portuguese power in South Africa all this was forgotten, and the very existence of the mines and monuments had faded from the memory of man until they were re-discovered by Adam Renders in 1868, and in 1871 again visited, and for the first time since De Barros, described by Dr. Carl Mauch and Thomas Baines.[7] But for the first thorough survey, at least, of the chief central group of buildings, 'Great Zimbabwe,' near Victoria, we are indebted to the distinguished archæologist, the late Theodore Bent, whose classical work, 'The Ruined Cities of Mashonaland' (London, 1892), was certainly the best and most comprehensive book of reference on the whole subject till the appearance of Messrs. Hall and Neal's sumptuous volume.

Like so many of his predecessors, and, we may now add, successors, Bent was strongly inclined to ascribe the ruins to the South Arabians, and so strongly impressed was he by their great age that he was prepared to assign them even to a pre-Sabaean epoch. 'The cumulative evidence is greatly in favour of the gold-diggers being of Arabian origin, *before the Sabaeo-Himyaritic period, in all probability*' (p. 186). But it is unnecessary, and indeed impossible, to go so far back; nor would Bent have written thus had he known, as we now know, that the South Arabian rock inscriptions partly interpreted by Mordtmann, Glaser, and others, carry the Sabaean and Minaean records back thousands of years before the new era, that is into late Neolithic times. Elsewhere he speaks more cautiously, and in fact retracts somewhat from the pre-eminence of Rhodesia as the chief gold-yielding land of antiquity: 'Here, near the east coast of Africa, far nearer to Arabia than India, and China, and other places, which they [the Sabaeans] were accustomed to visit, not only is there evidence of the extensive production of gold, but also evidence of a cult known to Arabia and Phœnicia alike, temples built on accurate mathematical principles, containing kindred objects of art, methods of producing gold known to have been employed in the ancient world, and evidence of a vast population devoted to the mining of gold.

'As to the vexed question of the land of Ophir, I do not feel that it is necessary to go into the arguments for and against here. Mashonaland may have been the land of Ophir or it may not; it may have been the land of Punt or it may not; Ophir and Punt may be identical, and both situated here, or they may be both elsewhere. There is not enough evidence, as far as I can see, to build up any theory on these points which will satisfy the more critical investigation to which subjects of this kind are submitted in the present day. All that we can satisfactorily establish is that from this country the ancient Arabians got a great deal of gold' (p. 194).

Since these words were penned much additional evidence has been produced by the more extensive and systematic surveys of the Rhodesian lands. But few serious attempts have been made to utilise this evidence in the solution of the difficult problems associated with the numerous pre-historic remains strewn over the region between the Zambesi and the Limpopo. The subject remains practically as it was left by Bent, and science still withholds a convincing reply to the questions—Who built these structures? who worked the surrounding gold-fields? Was this auriferous region—ancient Rhodesia, as we may now call it—the source whence David, Solomon, and Hiram, and the Queen of Sheba, and Arabia itself, drew their great store of the precious metals? Was this, in a word, the Ophir, or the Havilah, of Holy Writ?

Nevertheless some advance has been made; several wild hypotheses have been eliminated, and although one or two new claimants—notably the land of Punt and *north-east Arabia*—have somewhat unexpectedly made their appearance, the issues have been narrowed down to about five alternatives; the two just mentioned, *Madian* (*Midian*); *India*; *the Malay Peninsula, and Rhodesia.* No other *provenance* is any longer thinkable, and I propose here to produce fresh evidence, which should settle the question once for all in favour of

Rhodesia as the main source, and of the Himyarites and their Phœnician cousins as the chief forwarders, of the 'Gold of Ophir. At the same time it will be made clear that Ophir itself was not a gold-producing land, or, strictly speaking, a land at all, but a seaport, the chief Sabaean emporium on the south coast of Arabia, whither the precious metals and other exotics were forwarded and thence distributed over the eastern world.

Before leaving Rhodesia, to examine the claims of her several rivals, it will be convenient to briefly summarise the results of the researches that have been made in this field during the decade that has elapsed since the appearance of Bent's work.

One of the most intelligent of recent explorers in Rhodesia is certainly Mr. Franklin White, who has devoted his attention more especially to the Dhlo-Dhlo (Mambo) district about the head waters of the Inciza affluent of the Limpopo some fifty miles north-east of Bulawayo. The extensive group of ruins here discovered are fully described and richly illustrated in a paper communicated by the explorer to the 'Journal' of the Anthropological Institute for June 1901. The ground plan shows a main building roughly egg-shaped 350 by 200 feet, the longer axis running north-west and south-east, with several outer enclosures and isolated groups, covering altogether a very large area, and giving the impression that here was an extensive and strongly fortified agricultural settlement of the foreign treasure-seekers. The outer ramparts and inner walls, with narrow passages as labyrinthine as those of Bent's Zimbabye, are built of small granite blocks, those of the lower courses being of larger size, but all put together without any mortar as generally everywhere. Except where raised in tiers with intervening ledges or terraces, the walls seldom exceed 8 feet in height, but are diversified, as at Zimbabye, either with lines of different coloured stones, or with chequered or zigzag or chevron patterns, or else with alternating granite and red-banded iron-stone slabs. The better finished parts faced, not south-east as at Zimbabye, but north and north-west, where was also the main entrance. Nor did the explorer pick up any sacred emblems, such as those described by Bent, so that Mambo seems to have been rather a fortified station than a temple—'probably one of the chain of strongholds connected with the main route from the east coast,' that is, from 'Sofala Bay probably the port of entry . . . If the Phœnicians were the builders they may have abandoned the country in the same manner as the Romans left Britain when their mother country was in the last stages of its existence. If this theory is correct these ruins would be at least 2,300 years old.'

But these monumental remains, generally associated with mining operations, and obviously erected for the comfort and protection of the gold-hunters, were not confined to the Matabili, Mashona, and Manica lands, but also ranged southwards beyond the Limpopo into the Transvaal. Here Mr. J. M. Stuart tells us that he came upon numerous old workings, 'showing that centuries ago mining was practised on a most extensive scale, that vast quantities of ore had been worked, and that by engineers of a very high order. I found quarries, tunnels, shafts, adits, the remains of well-made roads, and also pits of ore on the site of these old roads, apparently ready to be put into waggons. This ore was piled up with as much regularity as if it had been placed for strict measurement, and it would seem as if these workings had been abandoned precipitously by the miners. I found, in one instance, that a gallery had been walled up with solid masonry. The native tribes, so far as I could ascertain by diligent inquiry, knew nothing as to who these ancient miners were, and have no traditions regarding them.'[8]

It appears from the careful researches of Messrs. Hall and Neal, spread over a considerable period, first recorded in the 'Bulawayo Chronicle,' and now embodied in the important volume referred to in my preface, that numerous remains of old workings, temples, and fortresses occur in the whole region between Panda-ma-Tenka and Tati, and thence eastwards nearly to the coast, or a total area of some 115,000 square miles. The mines, properly so called, thousands of which now lie buried beneath a dense vegetation, or have been obliterated by the action of the atmosphere and running waters, have an average depth of from 30 to 50 feet, some even exceeding 150 feet, and it is estimated that gold to the value of 32,000*l.* had been extracted from one particular working after making all allowance for wasteful methods of production. Despite a general uniformity of type, the associated structures present varying degrees of excellence in their design, execution, and ornamentation, and a close inspection certainly points at several successive occupations of the gold-fields since the erection of the oldest monuments which Messrs Hall and Neal think may date from some 1,100 years before the new era. We shall see that this estimate lies well within the limits of probability, and also that, as suggested by them, the first prospectors were almost certainly the Himyarites of South Arabia, followed later by the Phœnicians, and then, perhaps after a long interval, by the Moslem Arabs and the Portuguese.

These diligent investigators distinguish four recognised types of remains, as under:

1. Those of the first and best Zimbabye period, class, or type.
2. Those of the second, and somewhat inferior Zimbabye type.
3. Those of a late and distinctly decadent Zimbabye type.
4. Period when local races endeavoured, with ill success, to adopt the Zimbabye style of architecture.

They estimate that there are altogether very considerably over five hundred distinct sets of ancient ruins, all built during the three Zimbabye periods proper, as above defined, and presenting the recognised features of Zimbabye architecture. 'This ancient colony,' they remark, 'was evidently well organised, and was divided into districts, each district distinct from the others, each with its capital town, possessing its extensive and well-defined temple remains, and numerous gold-smelting furnaces. While scattered both near and far around each of these capital towns, whose massive ruins to-day resemble ruined cities, are scores of ruins of similar buildings, mostly without temples or traces of ancient smelting operations, which appear to have served as bases of supplies for the workers on the adjacent reefs or shed-gold areas, or as temporary treasure stores, or as refuge or defence against attacks of the savage negroid races who lived in these territories, and from whom the slaves were drawn, who toiled for their alien task-masters. . . .

'The chains of ancient forts which occupy isolated positions of great strategical value at long distances from gold-bearing areas, may be explained as forts protecting roads, each trending due east, not only connecting centres, but leading in a well-defined line, continued in Portuguese territory towards the port of Sofala. . . . The forts, some of which must have occupied several years in building, with their massive walls—some fifteen feet broad—their intricate entrances, and other characteristics of their architecture, prove that this people considered protection to be of primary importance.'

Amongst the capital towns, each surrounded by gold-reef districts, are mentioned :—

Dhlo-Dhlo for the Upper Insiza Valley.
'Mnukwana for South Belingwe.
Tuli for the Gwanda district.
Tati for Macloutsie and Shashi.
Martindela for Sabi.
Thabas Imamba for Lower Shangani.
Khami for Bulawayo district.

Many of the objects found amid the refuse, or

beneath the cemented floors of the buildings, are of quite exceptional interest to the anthropologist no less than to the historian and archæologist. Some of the more important are specified in the subjoined brief account of a number of ruins, all of which date wholly or in part from the first Zimbabye period, and have been personally inspected by Messrs. Hall and Neal, to whose courtesy I am indebted for their reproduction here.

The Check Ruins, at the head of a western tributary of the Bulye in Belingwe district ; massive walls $3\frac{1}{2}$ feet at base, highest still 12 feet, built on a system of curves ; entrance walls rounded, all with the usual 'batter back' both inside and outside ; show splendid workmanship in decoration equal to any work of this period ; material, granite blocks ; area at least half an acre ; are the only structures yet found that are completely covered both inside and outside with decoration of the check or chess-board pattern ; pannings of the soil yielded some fine gold and parts of solid gold ornaments.

The Mundie Ruins, a group of four ruins three miles west of the Mundie River in Belingwe district ; within a mile are very extensive old iron workings, extending a length of twenty miles ; massive walls of the same batter as the oldest portion of Great Zimbabye, and same excellent workmanship ; were the gold-smelting centre for this district ; built of granite blocks, with walls 7 feet at base, and 4 feet at present top, with check ornamentation ; yielded altogether 230 oz. of gold ornaments, beaten gold, gold nails, and gold bangles, some weighing 6 oz. each. With an ancient skeleton were found interred gold ornaments weighing 72 oz.

Charred remains of unburied people, indicating a fight and massacre of the old occupiers; also similar signs of conflict outside one group—unburied remains, torn gold wire bangles, scattered beads (one gold bead weighing 1 oz.). In the *débris* were heaps of gold crucibles, and portions of blow-pipes, and five gold-smelting furnaces sunk into the cemented floor, with cakes of gold still in the crucibles, and cakes of gold near the waist of some skeletons, as if carried in a pouch hanging on a belt.

The 'Mnukwana Ruins, in Belingwe district, contained skeletons of a woman and a child, and 14 oz. of gold ornaments buried with them; 'two very small baby's bangles were also found.' Here was a covered entrance like that of the Nuraghe ruin at Ortu in Sardinia, the only one of its kind yet found in Rhodesia. (See the Ortu group, as restored by Cima, in the Hon. A. Wilmot's 'Monomotapa,' p. 17.) Under a floor was much old pottery, with ashes and bones, gold beads and pellets, 'also a very large number of broken crucibles with gold in the flux, and portions of blow-pipes with gold on the nozzles'; also 'double-bells, gold rosette, with rising sun image, and copper ingot.' In all, 72 oz. of smelted gold 'certified to be of a higher standard value of gold than that in British gold coinage.' The rosette of beaten gold, with sun image embossed—a common object in phallic decoration—resembles the knobs on the sacred cone of the great Phœnician Temple of the Sun at Emesa in Syria; eleven of such rosettes were found at Great Zimbabye.

The Chum Ruins, on a kopje 200 yards from west

side of junction of Malema and Tuli rivers in Gwanda district ; have never been disturbed or re-occupied after abandonment by their first occupants ; contained large quantities of old gold crucibles, with gold showing in the flux. Under an ancient cemented floor was the skeleton of a man fully seven feet high, in perfect state of preservation. Of these remains the skull, an arm, and a shin bone (this over two feet in length), were brought to England by the Hon. Maurice Gifford, and in February 1896 handed to Professor Thane of University College, London, for examination. But nothing further appears to have been heard of them. Messrs. Hall and Neal inform us (p. 105) that 'at the Monastery Diamond Mine, in the Orange River Colony, have been found giant remains of great age ; but so little is known of this find that it would be impossible to connect that race with the giant remains found in Rhodesia. However, it is important to remember that the Prophet Isaiah refers to the people of Taba (the suggested first occupiers of this country) as of high stature. Strabo, Pliny, and Diodorus also speak of the Sabaeans being physically robust, while Professor Rawlinson states that the Phœnicians were rather over than below the average standard of height.'

The gold bangles round the ankles of the Rhodesian giant were of immense size, and with him were found altogether about sixteen ounces of plain gold ornaments. Under other cement floors were other skeleton remains, all with plain gold ornaments, besides two gold bangles, evidently belonging to a small child.

With the Chum giant should be compared the

remains of Hen Nekht, an Egyptian king of the Third Dynasty (about 4,000 B.C.) described by Mr. C. S. Myers in 'Man,' October 1901, p. 152. 'His stature probably exceeded 1,870 millimetres,' while the skull was unusually broad for an Egyptian. The broader-headed race, apparently represented by this king, 'is commonly thought to have arrived first in the Nile Valley at Koptos from Punt, a land sacred to the later Egyptians, the situation of which it is conjectured was near Somaliland and the opposite coast' (*ib.*). This 'opposite coast' is South Arabia, whence also came the Himyaritic settlers in Rhodesia.

The Thabas Imamba Ruins, on north-west side of the range of like name, overlooking the Hartley Hill Road 600 yards off; partly of the first Zimbabye period, but with later additions of the second, third, and fourth periods, as at Dhlo-Dhlo and Khami. This was evidently a very large gold-smelting centre, 'as portions of many thousands of gold crucibles and blow-pipes of the very oldest pattern, with gold still in the flux, were found under the present floors. There are indubitable evidences that a very great population resided at or near these ruins.' It is incidentally mentioned that the King Mombo, of the Mombo period, lived here, and was here flayed alive by the Ama-Swazi people.

The M'teleqwa Ruins, in the hills of like name, on the north bank of the Shangani, facing its confluence with the Longwe. Besides numerous later remains, one skeleton was found buried with 3 lb. of gold; also many crucibles of the very oldest make, showing visible old in flux, and gold bangles with the Zimbabye herring-bone pattern, which London experts state 'was stamped on with one complete and perfect stamp.'

Messrs. Hall and Neal's researches have since been largely supplemented by those of Dr. Carl Peters in the metalliferous district between the Lower Zambesi and the Sabi rivers. Full particulars have not yet been published, but from his summary report in the 'Times,' August 1, 1901, we learn that all the ruins along the east frontier of Mashonaland belong apparently to the early Zimbabye cultural epoch. 'Along the banks of the Mazoe and Ruenya, on the high plateaux of Inyanga and Melsetter, in Manicaland, and in the neighbourhood of the Sabi river, I have everywhere found the same type of ruins. Over the whole district the cyclopean wall is in the typical form of house building, and in some parts *whole cities of these buildings are lying open to the eye of the astonished explorer.* In North Inyanga I found *débris* of ancient settlements with a diameter of 1½ to 2 miles . . . His [Bent's] theory on the character of the Zimbabwe and other ruins is undoubtedly quite correct. The symbols of phallus worship can be found from the Zambesi down to the Sabi, and there can be no doubt that the old race which has left these ruins adhered to the cult of Baal and Astarte . . . All these facts lead me to believe that the ancient conquerors belonged to a Semitic race, and the repeated appearance of the name of Massapa, Umsapa, Rusapa, Sabi, &c., make it highly probable that they were Sabaeans, a race very nearly related to the Phœnicians.'

Notwithstanding these astonishing revelations, which should at once place Rhodesia in the forefront of all her rivals, she has, so to say, been ruled out of court on the assumption that she possesses neither the silver, nor the precious stones, nor the peacocks, mentioned in Scripture amongst the treasures brought from abroad (Ophir, Tharshish, or both) to the courts of David and Solomon.

To the silver difficulty the answer is that, although not yet re-discovered, silver does occur in Rhodesia, as was well known to the Portuguese. What says Dos Santos, who knew the country, and was one of the most trustworthy of all their writers? In his graphic account of Baretto's disastrous expedition to the metalliferous lands of the interior he writes: 'Although the country is rich in gold and silver mines, yet their contents are not easily obtained, for the Kaffirs are prohibited under penalty of death from showing the mines either to their neighbours or to foreigners.' And farther on: 'On his way to the silver mines at Chicona, Baretto was opposed by the King of Mongas, but defeated him in several battles.'[9]

A clue to the position of Chicona, which was reached and for a short time held by Baretto, is afforded by the statement that he soon after returned to Sena with the remainder of his forces. Sena is well known, and is still a Portuguese station, on the right bank of the Zambesi, some distance above the Shiré confluence, so that the silver mines will probably be found somewhere in Manicaland, which lies in Portuguese territory, just east of Rhodesia. Indeed, 'the Kingdom of Manica' is mentioned by Dos Santos, who tells us that it was reached by Baretto in a few days after occupying and destroying Zemboe, capital of the Juiteve (another title of the Monomotapa already referred to). Zemboe is, of course, de Barros' Symboac, that is, one of the numerous Zimbabyes or 'royal kraals' occurring everywhere in the auriferous districts. One is almost tempted to identify this Zemboe, destroyed by Baretto, with the ruined Zimbabye explored by Bent and Swan. That the Portuguese passed this way is evident enough from the statements both of White, who was informed that 'two small Portuguese cannon and a considerable quantity of silver articles, such as would be used by the Jesuit Fathers [should be Dominicans] who would accompany a Portuguese expedition,' were found at Dhlo-Dhlo (*loc. cit.*, p. 27), and of Stuart, who speaks of 'gunpowder' in the workings much farther south beyond the Limpopo. On the strength of this discovery, Stuart goes so far as 'to attribute these workings to the Portuguese, who are historically known to have had many trading possessions, and to have gained much gold in this section in the seventeenth century. My reason for so attributing these workings is that they were acquainted with the use of gunpowder.' He adds, however, that whoever were the miners, 'the fact remains, and is open to all who will visit the country, that mining on a very extensive scale was carried on by some nation in the past.' [10]

In this connection it should be noted that, if silver has not yet been re-discovered in Rhodesia, it is known to occur in the neighbouring Transvaal, where it has even been mined in recent times. The semi-official 'Statesman's Year-Book' for 1901, p. 226, informs us that 'working for silver, lead, and copper has been suspended since 1894.' 'Argentiferous lead mines' are also mentioned by Reclus (vol. xiii. p. 209 of my English version). Hence the 'silver, iron, and lead' mentioned by Ezekiel (xxvii. 12), amongst 'the multitude of all kind of riches,' sent by Tarshish to Tyre, might all have been procured in Rhodesia, since Tarshish itself was probably the present Sofala, as will be seen farther on. At v. 25 'the ships of Tarshish' are again referred to, although, strange to say, the 'Vulgate'—Jerome's 'Vulgate'—speaks here only of *naves maris* ('ships of the sea'); while the Tarshish of v. 12 is rendered *Carthaginenses* ('Carthaginians'), with a boldness which should be a comfort to those who still cling to the authorised version, despite all its admitted shortcomings. But this by the way; and merely to show how carefully even the plainest texts of Holy Writ have often to be handled.

Touching the precious stones, which are several times (1 Kings x.; 2 Chr. ix., &c.) mentioned in the closest connection with the gold brought to Solomon, but are assumed not to be found in Rhodesia, recent exploration shows, on the contrary, that they do occur, and are distributed over wide areas. The South Africa Chartered Company's Report for 1901 speaks of rubies—true rubies—though of small size and in small quantities. More detailed information is supplied by Bent's associate, Mr. Robert M. W. Swan, who re-visited Zimbabye and explored most of the Mashona and Matabili lands between 1891 and 1896, and points out that many of the numerous temples or wayside shrines discovered by him occur 'in some districts, as between Makloutsi and Tuli, where there seems to be no gold, but crystallised quartz minerals of many kinds, and sometimes these are of considerable beauty. Many of the precious stones mentioned in Revelation are found there.'[11] He suggests that the strangers came in quest of gems as well as of gold, and the gems themselves seem to correspond best with the 'bdellium and the onyx stone' associated with the gold of Havilah in Gen. ii. At least this bdellium (*bdolakh*) is translated 'krustallos' (rock crystal) by the Septuagint in Num. ix., where it is again mentioned.

Even if Eduard Glaser[12] and some others be right in regarding the Biblical bdellium, not as a gem, but a gum, this somewhat strained interpretation will not weaken, but rather strengthen, the claims of Rhodesia, gum copal of a fine quality having been a staple export from Sofala for time out of mind. The Septuagint (Alexandrian Greek) version has *anthrax* (lit. 'burning coal') for Gen. ii. 12, this term being also used for the carbuncle, ruby, and garnet, while in modern botany and commerce the expressions Indian, African, Sicilian, bdellium, have reference to 'false myrrh,' and various other kinds of gum resins. But the point need not detain us, since, either way, Rhodesia may exclaim 'our withers are unwrung.'

Before leaving Rhodesia, for the present, I should like to point out that amongst many of the surrounding Bantu peoples—the Ama-Tonga, for instance, about Delagoa Bay—a common word for the precious metal is *golc*, which, of course, is the English word *gold*. From the Arab and Swahili traders the Zulus have also in recent times adopted the Arabic word *mali* = 'money,' which they commonly use in the sense both of money and gold. But it is a remarkable fact that the Hottentot word for *gold* (not money) is *mari*, which by the normal interchange of *r* and *l* [13] is at once referable to the Arabic *mali*. In the languages of the lower races such terms are almost invariably loan-words from those of their more civilised neighbours, as, for instance, Finnish from Germanic, and Basque from Latin. But there are no Hottentots in Rhodesia, which, for probably over two thousand years, has been almost exclusively inhabited by Matukas, Makalakas, Bechuanas, and other Bantu peoples. How, then, did the Hottentots come by this Semitic word? To explain, we must go back far more than two thousand years, in fact, to the Himyaritic period, when the land can easily be shown to have been occupied by Hottentot tribes, who were afterwards driven by the intruding Bantus from the north to their present restricted domain in the south-west corner of the Continent. Lichtenstein was the first to show this, early in the eighteenth century, when he paid his famous visit to the Bechuanas, and nearly everywhere found the rivers and mountains and other permanent geographical features bearing Hottentot names. 'The former presence of the Hottentot-Bushman elements all over South Africa is proved by the geographical nomenclature of the regions now occupied by the intruding Bantus. Thus the names of most watercourses contain some dialectic form of the word *ib* (*ob*, *opo*, *eb*, *ap*, *iep*, &c.), which, in Hottentot, means *water* or *river*, as in *Gar-ib*= great water, (the Orange River); *Hyg-ap*, *Nos-ob*, *Mol-opo*, and others. The Wak-Wak of Edrisi's map (1154), which has so greatly puzzled historical geographers, may even be the Bushman *Kwa-Kwa*, showing the presence of these aborigines on the east coast, south of Sofala, whence, "long before the Portuguese circumnavigation of Africa, they were driven back by Kafir tribes."'[14]

The strong walls and ramparts raised by the Himyaritic miners at so many strategical points in the auriferous districts were therefore defensive works, not against the warlike Zulus, who came much later, but against the no less warlike Hottentots of those remote times. Some, perhaps, remembering the cringing nature of the present Griquas, Gonaquas, and other debased Hottentot-Dutch, or Hottentot-Kafir half-breeds, may be inclined to smile at the epithet here applied to the race. But that they were formerly an extremely bellicose people is well attested by the history of their full-blood Nama descendants throughout the nineteenth century in their present domain in Great Namaqua and Damara Lands. Thus is explained their Arabian word for *gold*, and in the explanation we have another strong argument for the vast antiquity of the foreign gold-hunters in ancient Rhodesia.

CHAPTER III

PETERS' 'EGYPT AND THE LAND OF PUNT'

IN his letter to the 'Times' above referred to, Dr. Peters advances various arguments in support of a new theory which seems at variance with his admission that many circumstances 'make it highly probable' that the builders of the monuments and workers of the mines 'were Sabaeans, a race very nearly related to the Phœnicians.' Relying on an Egyptian statuette which he found during his last expedition of 1899, and on the detailed record of the Egyptian Queen Hatshepsu's famous expedition to the land of Punt, he now infers that Ophir and Punt are one and the same land, and that this land was the present Rhodesia, which was first settled by the Egyptians.

'The country between the Zambesi and Sabi, I take it, is the Ophir of Solomon's epoch. I believe that the finds of my last expedition bring some evidence that Punt was indeed the same country [Rhodesia]. I believe that these mining districts were originally discovered by the Egyptians, and were an Egyptian colony; that the Phœnician epoch, of which many more relics are extant, probably followed the Egyptian, in fact, that the Phœnicians took the country some time between 1600 B.C. (Queen Hatsepu's expedition, and 1100 B.C.) King David's period.'

Here the Sabaeans seem to be excluded altogether, while their place is taken by the Egyptians as the first colonists, and Ophir identified with Punt, and Punt with Rhodesia, where the Phœnicians replaced the Egyptians. The statuette, to which extreme—in my opinion undue—importance was attached, was described as having 'hieroglyphics round its waist which have not been deciphered yet. I take it that this figure is about five thousand years old, and a representation of the goddess Isis.' But this view was frankly withdrawn after the figure had been submitted to Prof. Flinders Petrie, who reported on it as under :—

'Ushabté figure, probably of a courtier of Tahutmes III.

'Upper part of an Ushabté figure of pottery impressed in a mould. On the head is an elaborate wig, in each hand a scourge instead of a hoe. On the chest is the cartouche of Tahutmes III. (about 1450 B.C.). Three lines of inscription remain below, so much effaced that only the title Osiris can be seen; but there is no trace of a cartouche with it.

'The wig and the scourges in the hands point to this being a figure of the King himself; but his name cannot be traced in the lower inscription.

'As to the source, the figure is certainly genuinely ancient, and by its smell it has been buried in moist earth (not in an Egyptian tomb), and has not been kept long by an Arab. All this agrees with the account of its finding.'

This 'diagnosis' was taken by Dr. Peters as adding more weight to the conclusion he had drawn from such a find south of the Zambesi, than his own original opinion. 'That a statuette of Thotmes III. [of the eighteenth dynasty, and nephew of Queen Hatshepsu] should be found in South Africa is indeed very remarkable, and may well lead the attention of scholars to the re-opening of the Punt question' ('Times,' August 20, 1901).

Why I do not attach so much importance to this relic, whatsoever or whomsoever it was intended to represent, I explained in a letter to the 'Times' (August 15, 1901) pointing out that it could not possibly 'prove any direct Egyptian influence in South Africa. Such an object might in the course of 5,000 years find its way through a dozen channels—the Phœnician traders, for instance, or the Sabaeans, Minaeans, or Axumites—to any part of the east coast, and thence to the interior.'

Regarding the identification of Punt with Ophir, and its removal from South Arabia and Somaliland to Austral Africa, I have to point out, with the Rev. A. H. Sayce (a supreme authority in questions of this sort), that the Biblical Cush 'embraces not only the Ethiopia of the classical geographers, but also the south-western coast of Arabia and the opposite coast of Africa as well [Somaliland]. It thus corresponds to the land of PUN of the Egyptian monuments, as well as to the Kesh or Ethiopia [the Egyptian *Kesh* is generally regarded as the same as the Cush of Genesis]. It was inhabited for the most part by a white race, whose physical characters connect them with the Egyptians.'[15]

And, again: 'Egyptian tradition pointed to the "divine land" of Arabia Felix as that from which their principal deities had migrated. Hathor was the goddess of Pun, Ra had journeyed like the Phœnix from the Arabian land of spices. The "divine land" was Southern Arabia, the source of the sweet-smelling incense which was offered to the gods. It was also the source of the sacred trees which the Egyptians planted beside the temples of their deities. The migration of the deities implies the migration of their worshippers as well.' In the eighteenth dynasty the term Pun was extended from South Arabia to the opposite coast of Africa, to Somaliland. 'In colour, form, and features, the inhabitants of Pun resemble the inhabitants of Egypt' (*ib.* p. 92).

Here it has to be noticed that the original Punt was South Arabia (Arabia Felix, Yemen), whence the name was extended to Somaliland during the eighteenth dynasty, say, about 1700 B.C. Now, it must be asked, if this land of Punt was Ophir, and Ophir Rhodesia, did the Egyptians get their Ra and their Hathor, and their other deities, as well as their gold, from this Rhodesia, which, as above shown, was at that time still inhabited by Bushman-Hottentots? Did they come themselves with their *lares et penates* from that region? Were they Hottentots? I am here reminded that some early philologists, Dr. Richard Lepsius[16] amongst others, have discovered some connection between the Hottentot and Egyptian languages, and

Dr. Peters himself adduces this very assumption in support of his views. But the affinity has not been confirmed by later research, and all sound linguistic students now regard the gulf between the two languages as impassable. Egyptian is a member of the Hamitic family, which appears to have branched off ages ago from a common Hamito-Semitic stock; while Hottentot is a language *sui generis*, with no known structural or lexical kinship with any other.

But Dr. Peters places more reliance on the detailed account of Queen Hatshepsu's expedition as illustrated on the monuments. To this he calls my special attention, and writes me (August 19, 1901): 'The representations of the Punt expedition under Hatsepu [Hatshepsu] anyhow prove that it was an enterprise over sea (eight ships and one boat), and the freight there depicted may well have come from South Africa (particularly ivory and gold, also gums, leopard skins, and apes).' There was a giraffe also, and on the monuments is figured a ship, quite like an Arab dhow or large sambuk, being laden at the Punt coast with mghariot shrubs (frankincense), straight-backed oxen, men with loin-cloths, costly woods (especially ebony), cosmetics, dogs, slaves, besides the other wares mentioned by Dr. Peters. But all this proves nothing more than that this Punt lay on the African side, Arabia being excluded by the giraffe, oxen, ebony, and, presumably, the slaves. Hence Glaser (ii. 298) infers that Punt (Puna, Pano) may be the Opone of the 'Periplus Maris Erythraei,'[17] extending along the east Somali coast from Cape Guardafui to the equator, with a 'hinterland' stretching for any distance inland. But he at the same time warns us that it is not to be identified with Ophir, as was done by Brugsch, who was induced by the giraffe to transfer Ophir itself from Asia to East Africa.

As gold is mentioned amongst the exports from Punt, I may say that, in my opinion, the Egyptians seldom, if ever, sent out expeditions specially for the precious metal. They had plenty of it at home, and the famous mines in the Elba Hills on the Red Sea facing Jeddah were worked by the laborious crushing process for long ages. In this connection Dr. Peters' figurino, armed with a scourge in each hand, and picked up in a mining district, has a curious significance. The absence of a cartouche and other features seem to show that it does not represent a king, or any high official, but rather a superintendent of mines such as those of Elba, where Agatharkhides tells us that the hapless slaves—men, women, and children—were worked literally to death under the ganger's lash. The vivid description of their life-agony, which touched even the hard hearts of those days, is worth quoting: 'Those who are thus condemned to penal servitude, being very numerous, and all in fetters, are kept constantly at work both by day and night, without any repose, and are jealously guarded to prevent their escape; for they are watched by companies of barbarian soldiers who speak a language different from theirs, to prevent their winning of any of them over by friendly intercourse or appeals to their humanity. . . . Unkempt, untended as they are, without even a rag to hide their shame, the awful misery of these sufferers is a spectacle to move the hardest heart. None of them, whether sick or maimed or aged, not even weak women, meet with compassion or respite ; all are forced by blows to work without intermission until they expire under the hard treatment. So overpowering is their affliction, that they are ever anticipating worse evils in the future, and welcome death as a blessed change from life. The general superintendence of the mines is entrusted to the artificer who tests the stone, and he directs the workmen ; and the strongest in limb of those who are doomed to this hard lot break away the glittering marble with iron hammers—and that by main force in default of skill—and excavate subterranean passages, not indeed in straight lines, but following the cleavage in the gleaming rock. These workmen, as they pass their time in darkness owing to the turnings in these galleries, wear lamps attached to their foreheads.... On this they are unceasingly occupied under the lash of an exacting taskmaster. Then the children who are under age penetrate through the galleries into the chambers hollowed in the rock, and having laboriously thrown up the fallen pieces, convey them into the open, to a place set apart for the purpose outside the pit's mouth.' [18]

CHAPTER IV

BURTON'S 'GOLD MINES OF MIDIAN'

PASSING now from Lower Egypt across the Red Sea to Arabia, we come at once to the famous land of Madian (Midian), whose claims to be the auriferous Havilah, or perhaps the Ophir itself, of Scripture, were first seriously examined by the late Sir Richard Burton. He deals with the subject in two separate works—'The Gold Mines of Midian and the Ruined Midianitic Cities,' 1878, and 'The Land of Midian (Revisited),' 1879—and it will be noticed that the glowing title of the first is already toned down in the second. There is no longer anything about 'Gold Mines,' but only the plain 'Land of Midian,' and, truth to say, the erudite but somewhat erratic author seems at last to weary of his theme, leaving the impression that Midian, like so many other clamorous competitors, has failed to make good her claim. During the first visit, which, however, lasted only a fortnight, little or nothing was brought to light, and the explorer candidly admits that 'Midian is not included by Hebrew Holy Writ in auriferous Arabia, yet it has evidently supplied the precious metal in abundance, and it still deserves a place amongst the mining regions which in olden time made the peninsula 'Eudaemon' ('The Gold Mines,' p. 250). Where is the evidence of this? And in any case it may be asked, How could the products of Midian, a district in *Arabia Petraea*, help to confer the epithet of 'Eudaemon' (Felix, Happy) on the whole peninsula, to which it was never applied, but always restricted to the south-western uplands (Arabia Felix=Yemen)? Burton then appeals to his learned friend, Sprenger, who is asked to cast about for evidences of the occurrence of gold in the peninsula amongst the Greek, Roman, and Arab writers. But Sprenger finds little beyond vague statements, conjectures, strained etymologies, rumours and reports, and asks, in his best-known published work,[19] 'Are we to believe all these reports, the mere fancies of poets,' one of whom sings, with Oriental hyperbole, 'The Land of the Oqayl raineth gold'?

During his second expedition, Burton really did come upon two ancient gold workings, and it will be instructive to hear what he has to say about them. Of the first, at Umm el-Karayát at the Jebel el-Maru, he tells us that for the first time during the whole journey here were to be seen 'signs of systematic and civilised work. In many parts the hill has become a mere shell. We found on the near side a line of air-holes cut in the quartz rock, and preserving a rim, sunk like that of a sarcophagus, to receive a cover. Possible it was a precaution against the plunder which ruined Brazilian Gongo Soco.' One pit, 60 feet deep, 'had two main galleries both choked with rubbish. In other parts the surface was pitted with shallow basins, and a large depression showed the sinking of the hollowed crust.' Elsewhere 'specks, or rather *paillettes*, of gold were found lightly and loosely adhering to the *Maru*, so lightly, indeed, that they fell off when carelessly pocketed.' The search for relics yielded a few coarse stone grinding or crushing implements, and 'the split half of a ball of serpentine with depressions probably where held by finger and thumb. . . . There were no sign of wells or aqueducts, and the few furnaces were betrayed only by ashen heaps, thin scatters of scoriæ, and bits of flux—dark carbonate of lime.' 20

'At Umm el-Haráb [the second mining centre] we saw for the first time an open mine scientifically worked by the men of old. . . . They began by sinking a line of shafts across the summit, which may measure 120 yards. The intervening sections of the roof are now broken away; and a great yawning crevasse in the hill-top gives this saddle-back of bare cream-coloured rock the semblance of a "comb" or cresting reef.' A descent was made into the chasm, which apparently did not exceed 30 feet in depth, 'but in places the hollow sound of the hammer suggested profounder pits and wells. I should greatly doubt that such shallow sinking as this could have worked out any beyond the upper part of the vein. . . . The sloping roof has been defended from collapse by large pillars of the rock left standing out as in the old Egyptian quarries [There is a very fine coloured plate showing this]; it shows the clumsy but efficient practice that preceded timbering.'

Three kinds of quartz-crushing implements were found, one 'an admirable hand-mill of the compactest grey granite, smooth as glass and hard as iron. Around the pin-hole are raised and depressed concentric circles intended for ornament; and the "dishing" towards the rim is regular as if turned by machinery. We have seen as yet nothing like this work; nor shall we see anything superior to it' (*ib.* p. 175). This remark is explained by the fact that, although a great traveller, Burton had not visited the Zimbabye monuments. But it would be a waste of time to compare these poor remains of the half-barbarous North Arabian Ishmaelites with those of the cultured South Arabian Himyarites in Rhodesia.

Finding no place in North Arabia for the auriferous Havilah and Ophir of Scripture, Burton indulges in the conjecture that, instead of being a single mart, 'Ophir applies to several countries; that it means the "Red Land," an epithet equally fitted to Eastern Africa and to Western India; and that when the "ships of Tarshish and Ophir" are mentioned, the reference is to large vessels built for buffeting the stormy seas of the Farthest West and the Farthest East.' Thus in the 'Gold-Mines,' p. 262; and in the 'Land of Midian,' p. 239, he writes 'Whilst owning that one of the Ophirs or "Red Lands" lay in the modern Yemen, somewhere between Sheba and Havilah [*Khaulán*], I see no reason for concluding that this was the only Ophir. Had it been a single large emporium on the Red Sea, which collected the produce of Arabia and the exports of India and of West [read East] Africa, the traditional site could hardly have escaped the notice of the inquiring Arabian geographers of the Middle Ages. The ruins of a port would have been found, and we should not be compelled theoretically to postulate its existence.' The last passage has a peculiar significance in the present connection, for it will be seen in its proper place that this 'large emporium' and its 'ruins' have been found, and that very nearly where Burton looked for them. It will also be seen why this port of Ophir, 'somewhere between Sheba and Havilah,' escaped his notice as well as that of 'the inquiring Arabian geographers of the Middle Ages.'

Meanwhile attention may be called to the fact that the Ezion-Geber, where Solomon founded his naval station and built his 'ships of Tarshish,' stood near the head of the Gulf of Akaba on the coast of Midian. This district is therefore at once excluded, since it would be absurd to suppose that ships were built on the coast of Midian to reach the Havilah, Tarshish, or Ophir gold-fields in the Midian uplands.[21]

Thus Burton's 'Gold Mines of Midian' may now be dismissed, since it is obvious that the presumably inspired writer could not possibly have had Midian in his mind when in Genesis x. he suddenly introduced Havilah 'where there is gold.'

CHAPTER V

GLASER'S 'HAVILAH AND OPHIR'

My position would of course be greatly strengthened if Arabia itself could be as summarily dismissed as the Land of Midian on its north-west frontier, and if we could accept without further enquiry the current statements that there is no gold in the peninsula. But in point of fact the most contradictory views continue to prevail even amongst the greatest authorities on this very subject. Owing partly to the inaccessible nature of the land, and partly to the fanaticism of its nomad inhabitants, a great part of the interior still remains almost a *terra incognita*, so that a large field is left for conjecture. How hopelessly irreconcilable are the assertions of writers 'who ought to know' will be placed in a vivid light by the subjoined passages disposed side by side for convenience of comparison :—

Gold in Arabia

'It was the Arabians above all who lavished amongst the nations most of the gold occurring in ancient times.'—Sprenger, *op. cit.* p. 299.

'Here [Arabia] was a flourishing

No Gold in Arabia

'We have the authority of Niebuhr, that the precious metals are not found or known to exist in Arabia, which has no mines either of gold or silver.'—Crighton, *Hist. of Arabia*, ii. 403.

Gold in Arabia

gold export trade, and the seaports of the famous auriferous land of Ophir are to be sought nowhere but in East Arabia.'—Glaser, *op. cit.* p. 84.

And the 'Periplus' is quoted to prove that gold was actually exported to India itself (Barygaza, the present Baroche) from Ommana and other old Arabian seaports!—*Ib.* p. 80.

No Gold in Arabia

'Gold really never existed either in Arabia, or in any island along its coasts.' And, 'No gold, nor trace of gold-mines, has been discovered in Arabia.'—Smith's 'Dict. of the Bible,' *s.v. Ophir.*

No wonder that Glaser is himself perplexed, indeed driven almost to desperation, by such conflicting statements, exclaiming: 'It might almost seem as if on every one of my Himyaritic stones were chiselled the words *Lasciate ogni speranza voi che entrate*' (*ib.* p. 317). What follows is highly characteristic. When at the very gates of hell itself, he seems to see a glimmer of light which may lead him straight to the Earthly Paradise—the real Biblical Paradise, with its Pison, the golden Havilah and the rest—so 'away with all these hellish thoughts! Eden, thou fairest of all gifts of a kindly God, be thou greeted by me' (*ib.*).

This means in less ecstatic language that he really thinks he has found in North Central Arabia the long lost site of the terrestrial Paradise, in this now silent wilderness the very Pison—the Wady ed-Dawâsir—'that encompasseth the whole land of Havilah, where there is gold' (Gen. ii. 11); and hard by the land of Ophir with its havens and harbours, formerly indeed on the east, but lately (some 2,000 years ago) transferred to the west side of the neighbouring Persian Gulf. So at least I understand him to mean, for the argument, occupying many closely printed pages and supported by a boundless wealth of learning, is not easy to follow.

At first sight it is not clear why Ophir has to be located originally on the east and then shifted to the west side of the Gulf. But the reason is because in North or Central Arabia no word could be found etymologically connected with 'Ophir.' Hence it has to be sought farther east, where the *Apyron* mentioned by Diodorus and others 'will serve.' This Apyron-gold 'is evidently Ophiritic gold (Âpir=Ophir),' and when Ophir was forgotten a popular Greek etymology was found for Âpir.[22] Then referring to Lassen and Ritter's suggestion about *Abhîra*, east of the Indus, Glaser is disposed to conjecture that the names 'Ophir, Apira-ak, Apira, Khapira, Apirus, Cabirus, and Abhîra are identical' (p. 377). He adds: 'In this connection I recommend the philologists to enquire whether after all the Greek word ἤπειρος [the ordinary Greek term for continent, mainland], which in nearly all the passages of the "Periplus" should apparently be translated *shore*, *strand*, *coast*, is not somehow connected with the original *Apir*, *Ofir*, so that this word may strictly speaking correspond to our word "seaboard" (Kustenland).'

Carl Ritter also, before he finally decided for South Arabia, as we shall see, showed strong velleities for an Indian Ophir, influenced largely by these same etymologies. On this point Dr. Peters has some instructive remarks (*op. cit.* p. 50): 'Lassen and Ritter recognise the name Ophir in the people called Abhira at the mouth of the Indus. It was the nearest Indian coast for the Phœnicians, and here they could find stores of the goods of the north and the Himalayas, such as gold and bdellium and those of the south, such as sandalwood, &c. These Abhira, with many others of their tribes, which at the same time represented different castes, were also settled in the northern Punjab, but migrated thence in a southern direction, when Brahmanic [Hindu] colonies of the very earliest times still led a peaceful pastoral existence, and thence these Abhira first reached the coast. . . . They must therefore belong to the oldest Brahmanic tribes, descendants of Indian Aryas, and held sway in Solomon's time over that coast, so that the name Ophir must have been derived from them. Their migration to the mouth of the Indus must have taken place within the second millennium B.C.'

'Then Ritter proceeds with brilliant ingenuity, and with his vast knowledge which embraces the entire world, to furnish the proof that all the goods enumerated in the Ophir voyages were procurable at the mouths of the Indus. It is not necessary to our research to follow him in the details of his arguments. It is sufficient to state that Carl Ritter has proved that gold, silver, ivory, sandalwood, peacocks, and precious stones could be procured in the country of the ancient Abir [Abhira], and that, consequently, it could well have been the objective of the Ophir voyages. Nevertheless the question is not solved thereby, how the Jews could have paid the Abir for such enormous quantities of gold as those mentioned in the Old Testament.

'Ritter himself cannot explain how an agricultural country like Solomon's Israel could have paid a pastoral people by barter for such vast treasures. . . . I am personally convinced that mining enterprise was the main object of the Ophir voyage, if such vast stores of gold as 420 or 450 kikkars represented the result of a single expedition. . . . That the conjecture of the acquisition of these masses of gold by barter is a difficult question for the Indian theory is admitted by Ritter himself.'

In all this Dr. Peters stands on sure ground, and we shall see in the next chapter how the difficulty is overcome, and at the same time how impossible it is to formulate any convincing theory which leaves the old Rhodesian gold-workings out of consideration. But the Abhiras, who have troubled commentators for a long time, can be disposed of even in a more effective way. The name is not tribal or ethical at all, but purely social. *Abhir*, with its variant *Ahir*, is the designation of the widespread caste of cowherds, and those Abhirs spoken of by the Greek writers were merely the pastoral peoples who occupied and still occupy the rich grazing grounds about the banks of the Lower Indus, and thence towards the lower slopes of the Suleimân range. With Ophir the word had no more relation than had these rude grazers themselves with the costly wares of Solomon and Hiram.

According to the census of 1891 all the Ahirs numbered 3,918,000, and Mr. W. Crooke ('Tribes and Castes,' i. 50) tells us that this pastoral caste 'was formerly of much greater consideration in India than it is at present. In the Ramâyâna and Mahâbhârata the Abhîras in the west are spoken of; and in the Parânik Geography the country on the western coast of India from the Tâpti to Devagarh is called Abhîra, or *the region of cowherds* [not the land of gold, for indeed here is no gold]. When the Kattis arrived in Gujarât, in the eighth century, they found the greater part of the country in the occupation of the Ahîrs. The name of Asirgarh, which Farishta and Khizâna Amîra say is derived from Asa, Ahîr, shows that the tribe was of some importance in the Dakkhin [Dekkan] also, and there is no doubt that we have trace of the name in the *Abiria* of Ptolemy which he places above Patalene.' Amongst some of the sub-groups the name takes the form of *Aphariya* or *Aphiruja*, which comes even nearer to Ophir. But they were cowherds from all time, and according to the inexorable laws of caste could not traffic in gold or have any *commercial* relations at all with other peoples. They were stock-breeders then as now, and assuredly from them the 'merchants of Tarshish' could never have obtained the precious freights which they went to Ophir in quest of.

But Glaser, having got his Ophir by the help of the Abhirs from India down to the east, it was still necessary to transfer it to the west side of the Persian Gulf, since here alone were situated any well-known gold-exporting havens.

The 'Periplus' makes special mention of one—*Ommana*—with which may perhaps also be included *Apologos*, the Assyrian *Ubula*. But Ommana forwarded not only gold, but also *ivory*, which is not found in Arabia, while the Assyrian Ubula was never heard of till the time of Tiglath-pileser III., and Sargon III., say, about 700 B.C. Hence, if Ommana was really 'the chief staple for the export of gold' (p. 352), it could only have been under the later Assyrian rulers, who had spread their sway over the Persian Gulf and surrounding lands, but not originally (4000 to 1000 B.C.) when the Himyarites and their Phœnician successors dominated in the Indian Ocean, and drew their supplies from Rhodesia and not from Central Arabia, where there are no monuments or any old workings worth mentioning.

Nevertheless, Glaser has taken immense trouble to make good the claims of the Arabian Havilah, which he identifies with the Jemâma district, in and about the Wady ed-Dawâsir basin. As this appears to have been a somewhat narrow tract, it may be taken in a wider sense as the special gold land of the ancients, hence 'without doubt is to be identified with the Biblical Havilah, as I have done' (p. 350). Hence also Sprenger is wrong, Delitzsch is wrong, everybody is wrong, who looks for the golden Havilah anywhere except in Arabia.

But to arrive at this result some straining is needed. Thus 1 Sam. xv. 7: 'And Saul smote the Amalekites from Havilah until thou comest to Shur, that is over against Egypt,' not suiting the argument, *Havilah* is changed to *Hakilah*,[23] for the first time, as far as I know, by any commentator. The word itself is stated to be undoubtedly identical in part ('theilweise') with the *Mêlukha* of the cuneiform inscriptions which already yielded gold and *ushu* wood (*almuggim*) in the time of Gudïas (2800–3000 B.C.). But the identity does not require one term to be completely covered by the other. On the contrary, 'a section only of Havilah would appear

to have been called Mêlukha by the Assyrians. Magan was its seaboard, just as Ophir is to be taken as the seaboard of Havilah' (p. 326). How can this be, since nearly all Assyriologists take Magan to be the Sinai Peninsula, which lies not on the west side of the Persian Gulf, but at the head of the Red Sea? But in Gen. x. there are two Havilahs, one a son of Cush, therefore a Hamite (v. 7), the other a descendant of Joktan, therefore a Semite (v. 29). So now the first Cush is corrected with the Assyriologists to *Kash*, who before 2300 B.C. is met in Babylonia, and later (sporadically) in Elam (Persia). The Chaldaeans themselves [a Semitic tribe originally about the head of the Persian Gulf] were a people akin to these Kashites (Kashdu, Heb. Kasdîm). The Mahras also [whom we shall meet again in a most important connection] are a surviving remnant of the Kashite aborigines in Arabia (*ib.*). All are, therefore, presumably Semites, and the old difficulty of Hamitic Cushites and Semitic Cushes is thus got over. This is certainly better than Sayce's arrangement, who first tells
us that 'the ethnologist must be content to leave the sons **24**
of Noah to the historian or theologian,' and then proposes a scheme of classification, in which the Egyptian Hamites are grouped with the Phœnician and South Arabian Semites, all as Hamites. His explanation of the two Havilahs is also remarkable: 'The South Arabian tribe of Sheba spread far to the north through the sandy desert of Havilah [this is Glaser's Paradise!], and founded a kingdom which came into conflict with Assyria in the days of Tiglath-pileser and Sargon. It is consequently named twice, once as a people of the south under the

head of HAM, once as a people of the centre, under the head of SHEM' (p. 41). This is like saying that when the English went to New England and came into conflict with the Algonquins they became Red Indians. But Sayce is not a systematist, and he redeems this lapse by elsewhere frankly stating that Gen. x. 'lays no claim to be an ethnological record. On the contrary, it tells us as plainly as language can speak that with ethnology and the ethnologist it has nothing to do. There may be ethnological documents in the Bible, but the tenth chapter of Genesis is not one of them' (p. 68). And elsewhere: 'It is descriptive merely . . . Its main purpose is *geographical*' (pp. 39–40).

Returning to Glaser, a closer identification of the Pison is attempted by bringing it into connection with the *Faishan* of the Sabaean rock inscriptions, reference being given to his own numbers, 529, 926, &c. The word, he tells us—and he is admittedly the first of modern Himyaritic scholars—means 'depression, wilderness, or the like,' answering to the later (Koranic) Arabic *el faish*. 'We are therefore justified in connecting the district of *Faishân* mentioned by Hamdânî [an Arabian geographer] with the Pison of the Bible.' This seems a tremendous inference to be drawn from such slight premisses. But if the identification be accepted, it must vitiate the whole of the argument, for nobody will ever believe that the Pison of the Earthly Paradise was a mere 'depression, wilderness, or the like.' But elsewhere (p. 323) he refers to three other possible Havilahs, none of which could, however, have been 'compassed' by the Biblical Pison, which he makes to rise in Central Arabia and flow to the Euphrates, as indeed the Wady ed-Dawâsir occasionally does during exceptionally heavy freshets. One is Strabo's *Khaulotae*, presumably, he thinks, on the Persian Gulf, where Niebuhr knew of a *Huâle* (*Huwaila*), a station of the Benû-Khâlid tribe about the Bahrein Archipelago and El Ahsa But Pison could not have come this way, as it would thus have reached the coast far to the south of the Euphrates, of which Glaser assumes it to have been an affluent. Moreover, we have above seen that Burton has placed this Havilah (*Khaulán*=*Khaulotae*) not in the east, but in the west or south-west of Arabia, somewhere near 'the modern Yemen.' With etymologies you can do anything.

The second is a Kushite Havilah on the Sinus Avalites [or Abalites], that is, the Arabian Sea, but on the south side, somewhere about the present Zeila, on the North Somali coast. This brings us back to the African section of the land of Punt, which has already been disposed of.

Lastly, the third golden Havilah is 'in India, or God knows where. As we see, Havilah is sought from Africa to India, and from the Persian Gulf to Syria' (*op. cit.* p. 323). Before following it to India I may point out that the exceedingly elastic Abhîra, equated tentatively by Ritter and Glaser with *Apir*, *Ophir*, &c., already brings us to the very threshold, or even to the borderland itself, of that region. Pliny's Abhîra, the Abiria of the 'Periplus,' was a people or a land lying north of Syrastrene (Kachh) to the east of the Indus, that is to say, in the present Punjâb, answering to the Indo-Scythia of the Greeks, consequently well within the north-west frontier of India proper. Hence, to support his hypothesis, Glaser is obliged to say that 'if we include Abhîra also still in Ophirland, then this region stretched even still further away [to the east] beyond the Indus' (p. 378). But this is leaving his Ophir—a section of the west side of the Persian Gulf—far behind him, and the adjacent Havilah of Central Arabia should consistently be given up in favour of the far more plausible and long-established claims of the west coast of India,

> and thence
> To Agra and Lahor of Great Mogul,
> Down to the Golden Chersonese.
>
> *Paradise Lost*, xi.

CHAPTER VI

LASSEN'S 'INDIA AND THE GOLDEN CHERSONESE'

NOTHING has caused commentators greater trouble than the intrusive initial 'S' in the Septuagint variants of Ophir—*Sophir*, *Sopheira*, &c.,[25] to which may be added the *Sophira* of Josephus ('Antiq. Jud.' viii.). But the explanation is obvious enough. When in later (Alexandrian) times the search began for Ophir which still continues, it became identified with the here associated Sephar; and when Sephar itself could nowhere be found it was relegated to India, because somebody remembered that it was 'a mount of the east.' Thus it came about that Ophir became Sophir, and Sophir became India, and so firmly established was this belief that when the Moslem Arabs began to translate the Bible under Greek influences they also rendered Sophir 'El Hend' (India), as in 1 Kings ix. 28, and Isaiah xiii. 12. Then came the modern commentators, who, with the development of linguistic studies and the establishment of the Indo-European (Aryan) and Semitic families of languages, were able, the illustrious Lassen leading, to show, apparently by conclusive philological reasoning, that the Hebrew names of the ivory, the apes, and peacocks, brought with other exotics by the fleets of Hiram to his friend Solomon, were traceable to an Indian (Aryan) source, and that these commodities therefore came from India, which was thus again identified with Ophir, Sophir, and Sephar.

In the Hebrew texts (1 Kings x. 22, and 2 Chron. ix. 21) the word for ivory is *shen-habbim*, 'tooth of elephants,' where the stem *hab* is equated with the Sanskrit *ibha*, meaning 'elephant.' But the Hebrews did not need to go to India for this word, which lay close at hand in the Egyptian *ābu*,[26] and, as all know, there were elephants in Africa as well as in India. The next term *Qophim*, 'apes,' is in the same way traced to the Sans. *kapi*, although, according to Pierret, Egyptian again yields a nearer form, *ḳeften*, at least for a particular kind of ape. It is right, however, to say that Pierret is not regarded as a very good authority, and Dr. Budge tells me that the more general Egyptian term for ape is *ḳân*. This, of course, gives an impossible derivation ; but on the other hand the Sans. *kăpĭ* is not much better, both vowels being short, whereas in the Heb. both are long (*Qōphīm*), while the initials differ, the Sans. *k* being the Heb. כ, not the ק (q) of *Qophim*.

Lastly *Tukkiim*, translated 'peacocks,' is referred to the Sans. *çikhi* through the Malayalim *togei* or *toghai*, and Tamil *tokei*. But despite the great authority of Lassen, I reject absolutely this admittedly strained derivation, and until better informed deny that so early as the time of Solomon the Aryan invaders of India had already ranged southwards far enough to influence the Tamils and other peoples of the Dekkan, all of whom spoke and still speak dialects of Dravidian, a language fundamentally distinct from Sanskrit.

Indeed the Sanskrit derivation appears to be now given up, and later commentators go no further than the above-quoted Dravidian forms to explain *tukkiim*. Others have suggested that the bird was not a peacock at all, notwithstanding the Septuagint *taós* and the Vulgate *pavo*, terms which themselves present no little difficulty, but have no direct bearing on the present question. Dr. Peters quotes an African word, *tukka*, 'guinea fowl,' giving Carl Ritter as his authority, and asks, 'Was it not possibly the guinea-fowl that the writers of *Kings* and *Chronicles* meant by their *Tukki*? The flesh of these birds is delicious, and if their plumage is less gorgeous than that of the peacock, it nevertheless is a very handsome bird. In any case even the guinea-fowl must have been a rarity for Solomon and his court, well worth transport from a distance.' [27]

I have not, however, been able myself to 'locate' Ritter's *tukka* in any modern African language, but would ask of what avail are a hundred such feeble etymologies against the strong walls and the remains of the innumerable workings which are strewn broadcast over the auriferous Rhodesian plateau, and are found nowhere else to anything like the same extent either in India or in Arabia or in any other part of the ancient world? I say Arabia, because the peninsula is often referred to in this connection, and rightly so, though not in the sense intended by the expounders of the sacred texts.

It might perhaps be suggested that the absence in India of Himyaritic monuments and gold workings, such as those of Rhodesia, might be explained on the supposition that in India the gold, with the ivory and peacocks, was procured from the civilised Dravidian natives in the ordinary course of trade, and not mined by the Himyarites (and Phœnicians) themselves. There are, for instance, no distinctly Phœnician monuments, and but few traces of the presence of these ancient seafarers either in Spain, whence they certainly drew much silver if not gold, or in Britain, whence also they certainly drew much tin. Why then need such monuments be looked for in India, where there is at least some gold, although the island of Sardinia, where there is no gold, is thickly strewn with the so-called *nuraghe*, which the Hon. A. Wilmot has gone far to prove were undoubtedly Phœnician structures?[28] The presence or absence, they argue, of Semitic monuments—temples, towers, strongholds and the like—might therefore appear to prove nothing at all in the present connection, unless, indeed, these buildings were also directly associated with gold workings and ancient mining centres, as is the case, the objectors should add, in Rhodesia, but nowhere else, or at least nowhere else to anything like the same extent.

Here arises a most important consideration. In the preface to 'The Ancient Ruins of Rhodesia' Messrs. Hall and Neal remark that 'The undoubted fact of many millions of pounds sterling present value of gold having been extracted by the ancients from these territories—some accredited authorities placing the value, on a conservative estimate of portions of the gold-bearing districts only, at over 75,000,000*l.*—during the same period, which covers the period during which Biblical references are made to the gold of Ophir, and the admission by authorities that no part of the known world, India included, yields such overwhelming evidences of extensive, continuous, and successful ancient gold-mining operations having been carried on as are found on every hand in Rhodesia, leaves much substantial argument to be disposed of by the opponents of the theory.' The very first difficulty they will have to grapple with is this prodigious amount of gold supplied to the courts of David and Solomon. How could it have been procured by legitimate trade from the more or less civilised natives of India? What commodities had such a comparatively poor country as Palestine, North Arabia included, to offer in exchange? 'Milk and honey' are all very well as a poetical expression. But it is obvious that neither these, nor the olives and grapes, nor the roses of Sharon, nor the quails and manna of the wilderness, could serve the purposes even of a low-grade barter trade with the Dravidians of the Dekkan. The gold was clearly not obtained in the days of Solomon and Hiram by purchase and exchange. It must necessarily have been taken forcibly, mined in fact with slave labour under the sjambok, by the 'merchants of Tharshish' themselves. It was therefore procured, not from India, already thickly inhabited by settled populations, but from some such auriferous region as

Rhodesia, which, as we have seen, was at that time occupied by savage but warlike Hottentot hordes, who supplied the 'raw material' for working the diggings, and at the same time required to be warded off by such stout walls and defensive works as those of the Rhodesian Zimbabyes.

Now there are no evidences of slave labour in India, where indeed it is excluded by the caste system, by which the social classes are once for all defined and determined. Once a *dhobi* (washerman) always a *dhobi*, once a coolie always a coolie, and coolie (Hindi *kuli*) means a labourer, whose service could be obtained only by a contract based on the principle of payment for value received. Not so in ancient Rhodesia, where Hall and Neal (p. 181) have discovered in the Inyanga district the very 'slave pits' in which the native captives condemned to work in the mines were kept in confinement. On this subject 'Rhodesia,' October 19, 1901, reproduces from the 'Bulawayo Chronicle' a communication recently made to that paper by Mr. McDonald: 'One of the most interesting features of the Inyanga district is the extraordinary pits with which the whole country is filled; every two or three hundred yards one comes across them. Roughly, their depth is about 20 feet; they are circular in shape, and have a diameter of from 12 to 15 feet. They are now largely filled up with soil, trees of considerable size growing in them, showing that the pits belong to some long past age. The walls are built up with solid granite blocks, which are roughly dressed, and the construction is finished off in a workmanlike manner, better than any dry building of the present day. No mortar has been used in the building. An entrance to the pits was effected, at a point some distance from the edge of the pit, by a sloping gangway or path, very narrow, which when it disappeared underground was also walled over with granite blocks. It is like a small tunnel. The idea that they were slave pits emanates from the style of the entrance, and it is true that once in there the slaves could not climb up the walls and escape, and one guard could hold the place quite easily. Everything points to the fact that at one time the district carried a teeming population; the whole country has been tilled, and though there is a great depth of soil, the first 18 inches has not recovered its original fertility, it having been exhausted by these bygone agriculturists. It is only by going two or three feet below the surface that you encounter soil capable of producing good crops. This naturally took the local farmers some time to discover; but now they are aware of the fact, they are raising capital crops—better forage than I have seen anywhere else in Rhodesia. As you come from Inyanga along the present road, for some 40 miles you pass through a chain of old forts, some of which are in a good state of preservation. These forts are like the system of blockhouses along the railway lines in the newly conquered territories, but are much closer together, in many cases being within a hundred yards of each other, while I am safe in saying none are more than 300 yards apart. As you approach the Mashonaland railway, the line of forts breaks away apparently either towards Zimbabwe or the Sabi river, and I have no doubt that if followed up they would probably be found to continue to Sofala, or whatever port on the east coast was frequented by the traders, or was the outlet for the produce of the country. At many of the forts were what appeared to be ruins of what must, I think, have been watch-towers; at any rate, they are very similar to the famous Round Towers found in different parts of Ireland.'

On the other hand, there were certainly fortresses and towers and other places of refuge in Sardinia; but this was because the Phœnician intruders wished to settle permanently in that central land of the Mediterranean, and had to defend themselves against the aborigines, by whom it had been occupied since early Neolithic times. It may be incidentally remarked that the Sardinian structures, as illustrated in Wilmot's book, show quite an extraordinary resemblance to some of the more typical Zimbabye monuments. It can no longer be doubted that they are of Semitic, most probably Phœnician, origin, and would thus tend to show that, as I hold, the Himyarites were succeeded by the Phœnicians in Rhodesia. There are even 'Tombs of the Giants,' as if for the ancient 'Sons of Anak,' such as we have already met in Egypt and Mashonaland. Mr. Wilmot concludes that 'the builders of the Zimbabwes in South-east Africa and of the Nuraghes in Sardinia were Nature-worshippers of the early Phœnician cult, when stone-worship was one of the leading features of that religion. Certainly we cannot be wrong in concluding that the oldest of the Zimbabwes of south-eastern Africa were erected before the ninth century B.C. There is little doubt that some of them existed when

Hiram, King of Tyre, obtained gold for the Temple of Solomon ; and we shall see that ancient Monomotapa was probably one of the Ophirs referred to in the Sacred Scriptures.' Then it is added in a note that 'Ophir was a generic title for a rich commercial country, used in the same way as "Tarshish." The latter name we know was given to more than one place, as there may have been an Indian as well as an African Ophir. The former would be on the Malabar coast' (*ib.* p. 37).

Thus we are brought back to India, where there are two or three places on the west coast, such as the Barygaza, Muziris, and Nelkynda of the 'Periplus,' which might, in a very restricted sense, be regarded as such 'Ophirs.' Barygaza especially, which gave its name to the Gulf of Barygaza, that is, the present Gulf of Cambay, fulfils most of the conditions fairly well. It was the most famous harbour on this seaboard, at least in the first century of the new era, and probably long before, and among its exports the 'Periplus' (§ 49) mentions precious stones, perfumes, ivory, cottons, and silk. But again there is no gold, and without gold, and gold in abundance, there is no true Ophir—the 'Zarafshan,' or 'Gold Distributor,' as the Persians might call it, in a pre-eminent sense.

And so, like the Conquistadores in quest of their Eldorado in the West, we must still on, on, 'down to the Golden Chersonese,' the veritable 'Ultima Thule' of the Eastern world. And what shall we find here? At first one is startled to meet with a 'Mount Ophir' near the extremity of the Malay Peninsula, a region rightly identified with this very Golden Chersonese. But with the explanation comes the disenchantment, for we presently learn that this is only the Gunong-Ledang of the Malays, which lies in the moderately auriferous principality of Moar, about eighteen miles north-east of the city of Malacca, and was re-christened 'Ophir' in 1511 A.D., when that flourishing seaport was captured by the Portuguese pioneers, everywhere in search of the land whence Solomon drew his treasures. In the same way they discovered another 'Mount Ophir' in the Gunong-Pasaman, an extinct volcano about eighty miles north-west of Padang in Sumatra, and for all this they had the precedent of Columbus himself, who similarly reported to his royal patron the discovery of a golden Ophir in the New World (Hispaniola).

There is certainly a good deal of gold in the Malay Peninsula, which even now has a yearly output of from 80,000*l.* to 100,000*l.* Nevertheless, it does not appear to have been known to the Greeks as the Golden Chersonesus till the time of Ptolemy, who must have received it from Hindu sources. At least the Greek expression *χρυσὴ Χερσόνησος* is a literal translation of the Prakrit (late Sanskrit) *Suvarnabhumi*, the 'Gold Land,' which is applied in the Buddhist writings to a section of the east coast of the Bay of Bengal, that is the *Tanah-Malayu* ('Malayland') of the natives. M. Vivien de St.-Martin, who has made a special study of this subject, has come to the conclusion that *Cattigara*, farthest point of Graeco-Roman navigation in the extreme east, must have occupied the site of the present Singapore, which is itself a Hindu word, meaning the 'Lion City.' Here the Chinese had already, as they still have, settlements and trading establishments, and hither the Buddhist missionaries had penetrated at an early date, probably in the second century B.C. Thus may be explained the 'silk,' which was above mentioned as amongst the exports from Barygaza in the first century of the new era, that is 200 or 300 years before it is commonly supposed to have reached the West by the overland route from China.

The author of the 'Periplus' appears to have got no farther than the *Comori Promontorium*, doubtless Cape Comorin, and perhaps Taprobane (Ceylon), if so far. 'He states that opposite the island of Taprobane the coast trends towards the east, after which it pursues for a time a northerly course, and then once more turns eastward towards the mouths of the Ganges. Beyond that point lay a district called *Chryse*, and an island of the same name' (Tozer, p. 281). This is, of course, Ptolemy's Golden Chersonese, of which he evidently had but the vaguest information at second or third hand. The reference to gold occurs in § 63: 'They say that gold workings are also found in those lands; near the river itself is an island in the ocean [peninsulas were often called islands, especially by the Arabs], the last land of the inhabited world towards the east . . . which is called Chryse.'

Yet it is argued that the Malay Peninsula must have been well known to the ancients long before this time, because of the tin, of which it always has been one of the chief storehouses, and tin was already known to Homer.[29] It is even pointed out that *kassiteros*, the Greek for tin, is the Sanskrit *kastira*, whence also the Arabic *kasdir*. Here is another etymological pitfall, which has set the learned world astray for I scarcely know how many generations. In any case, it is now well ascertained that *kastira* does not occur till comparatively late times in Sanskrit, that in point of fact it is a loan word from the Greek, as is also the Arabic. There are other reasons, which I cannot enter into here, for believing that very little tin came from the East till relatively modern times, and that the chief sources of the supply during the bronze and early historic ages were Spain and Britain. 'On the whole, the most probable explanation, and that which agrees best with what we know of the concealment practised by the Phœnicians in respect of commerce, is this—that while the trade in tin continued to be their monopoly [from about 2500 to 300 B.C.], all that the Greeks learnt concerning its origin was that it was found in islands in the northern sea—by which Britain, together with the islets of its coast, or perhaps Ireland, are vaguely meant' (Tozer, p. 38).

We thus see that our Ophir cannot be sought in India proper, and still less in the ultra-Indian Golden Chersonese, which was scarcely known to the western peoples till about the dawn of the new era. Nevertheless Glaser, unable, like the Alexandrian Greeks, to find a resting place for Sephar in Arabia, suggests that this borderland of the Yoktanides [Himyarites] need not be in Arabia at all, and future research may show that it may be found somewhere east of the Persian Gulf, 'perhaps even in India, where there are places such as *Suppar*, with names sounding like *Sephar*' (p. 368). After this, one does not see why the philologists, B. Moritz,[30] for instance, amongst others, need have been so hard on Dr. Peters for venturing to suggest some possible relation between Sephar, Sophir, Sophera, Ophir, and Afir = Africa itself. 'The unprejudiced reader,' says this writer, 'will concede that the sequence Ophir, Afir, Afer, Africus, Terra Africa, and Africa is more than ordinarily suggestive. . . . I submit that my derivation of the name Africa from the ancient Hebrew word Ophir *is* the derivation which has been sought for centuries' (*op. cit.* p. 68). Then Africa being thus known as a whole, as a continent, to the Phœnicians, when we are told that they 'went to Ophir for gold,' we are to understand that this meant they went to Africa, so that Ophir is Africa (p. 85). The conclusion seems a little vague, and so far wide of the mark. Again, the root of the word means *red*, so that Africa is the 'Red-land,' analogous to the name Albion, which its chalk cliffs have conferred on England (p. 71). Ophir, he thinks, wrongly, as we shall see, also means red, as does Afar, the national name of the Hamitic Danakil people, between Abyssinia and the Red Sea. From Ophirland = the Latin Terra Africa, and from the Afar people we get, by the aid of an Egyptian prefix (p. 91), to Sofala, which is pronounced Sofara in the interior, and of course to this conclusion I cannot personally have any objection, because I want to get there also, only by a different route. Meanwhile Roland carries us back to India, because Upara or Uppara and the Malabar Supora were also called Sofala, and thus we have two Sofalas facing each other on opposite sides of the Indian Ocean, an *embarras de richesses*, which is got rid of by the assumption that the African Sofala is the original (p. 92). Here again I should be glad to agree, only I am reminded by Glaser, who on a point of this sort is the final appeal, that Ophir, once supposed to mean *riches*, and then *red*, cannot mean either. Burton had already rejected *riches*, but stood by *red*, because Seetzen had shown 'conclusively' that Ophir, the 'Red Land,' was in Southern Arabia. Now *red* also must go, because Glaser shows still more conclusively that the Mahra [South Arabian] *A'fur* has nothing whatsoever to do with 'red,' which, as he ascertained in Arabia, is uttered with the guttural *'ain*, and not with the *alef* (*a*),[31] and the interchange of these two letters in the Semitic languages is almost impossible. Thus collapses a whole continent; whatever 'Africa' may mean, it has no obvious connection with Ophir, and we have to begin all over again. Such is usually the outcome of arguments based on etymologies, to which I personally attach but slight weight, unless supported by collateral evidence. In fact, they may lead anywhere or nowhere, and while seldom settling anything, often tend to upset everything. Elsewhere, referring to electricity, I have spoken of the Protean phases of this new-born science, and the expression is too often equally applicable to the new-born science of philology, with its astonishing vagaries, possible and impossible transformations. So also the Indian Suppar, Supora, Sofala, Uppara, Upara, fall through, and we have to return to Arabia in search of the Ophir and the true Sephar, 'a mount of the east.'

CHAPTER VII

RITTER'S BIBLICAL OPHIR—THE FRANKINCENSE LAND

HAVING so often to differ from Dr. Peters in points of detail, I am all the more pleased to find myself in full agreement with his statement that one of the chief theories regarding the geographical position of Ophir—a theory 'adopted by men like Edrisi, Abulfeda, Bochart, Niebuhr, Gesenius, Vincent, Gosselin, Volney, Seetzen, Rosenmueller, Keil, and more recently by Soetbeer [and let me add Burton and Ritter]—looks for Ophir in Southern Arabia. There are indeed good reasons for it' (*op. cit.* p. 31). Were it merely a question of authority, the appeal to Ritter alone might almost suffice to close the controversy, just as Augustine closed the Donatist controversy by the famous exclamation, 'Petrus locutus est, causa finita est!'

But historic and scientific questions can no longer be settled by the *ipse dixit* of anyone, and all such matters have to be decided on their merits, as tested by intrinsic and extrinsic evidence. Where, however, absolute certainty is unattainable, that view is now generally accepted, at least as a working hypothesis, which is found to harmonise best with most of the known conditions of the problem. I hope now to be able to show that the position above assigned to Ophir is more than a working hypothesis, acquires, in fact, a certain finality from the fresh evidence brought to bear on the subject since it was discussed by the above-mentioned authorities.

As far as I can discover, the first clear pronouncement regarding the actual site of the Biblical Ophir is due to Carl Ritter, one of the greatest, if not the greatest, geographer of the nineteenth century. He writes: 'Since the dwellings of the children of Joktan are even geographically determined between Mesa or Mesha (Musa) unto Sephar (the early *Σαπφάρα*, the later Dhafar, Dhofar at Mirbat in the frankincense land, now the Isphor of the aborigines, by the "mount of the east," the present high incense range called Faguer in the Ehkili language), the very locality of the twelve tribes of the Joktanides, amongst whom is Ophir, is identified. Remarkably enough it coincides with the land of Mahra, east of Hadramaut proper, with the frankincense land renowned from the remotest times. But here was situated Dhofar (Zaphar, also Doffir, as Niebuhr heard it called), the ancient residence of the Himyaritic kings, during the most flourishing period of this State. It is still called "civitas eximia" in Ammianus Marcellinus (xxiii. 6, 47), and at the same time held from the earliest ages the position of an emporium for the great interchange of merchandise throughout the wide east, and might be at once indicated as that [Biblical] Ophir.'[32]

Here are concentrated in a few lines most of the elements required for the solution of the problem. The rest will now be supplied.

It has already been suggested that there is good reason to believe that the Biblical Mount Sephar should be brought back from India, where it is unknown, to South Arabia, where Ritter in the above-quoted passage identifies it with 'the mount of the east' in the present 'Dhafar or Dhofar at Mirbat.' Since Ritter's time this very district has been visited by several travellers such as Fresnel, Halévy, and Glaser, and was more particularly the scene of the very last archaeological labours of Theodore Bent, who in 1894–5 carefully surveyed most of the seaboard, which he justly describes as one of the most remarkable regions of the whole peninsula.[33] For us it is more than remarkable; it is of absorbing interest in the present enquiry, being the very district where, as had already been anticipated by Ritter, the Ophir of Scripture is assuredly to be sought.

Dhofar, as Bent tells us, forms a sort of oasis, an extremely rich level alluvial plain, extending some sixty miles along the coast a little to the west of the Kuria Muria Islands, and cut off by the Gara range from the sandy wastes of Hadhramout. Here still flourish both the myrrh and the frankincense shrub, which have constituted the chief industry of the inhabitants for thousands of years. Right up to the summit of the limestone range the slopes are honeycombed with great cavernous recesses, in which the natives dwell, as of old with their flocks and herds. Many of these natives are members of the Mahri nation, and still speak Ehkili, a dialect of the ancient Himyaritic language, which has elsewhere been superseded by the far more modern Arabic of the Koran, and is now known mainly from the innumerable rock inscriptions strewn over the whole region between Dhofar and the Red Sea.

According to Bent, Dhofar is the libaniferous land of the ancients, whose chief cities were Manteion Artemidos ('Oracle of Artemis') and Abyssopolis, near Moscha, or the Portus Nobilis, where the merchant ships plying on the Indian Ocean sought shelter during the prevalence of the monsoons. Here are scattered the remains of many temples, tombs, and public buildings, while the columns still standing 'form an interesting link which connects these ruins architecturally with the other ruined sites of the Sabaean world.' The columns at Axum in Abyssinia, at Adulis on the Red Sea, and at Maraiaba, ancient capital of Arabia Felix (Yemen), are all of the same type, and indubitably establish the Sabaean origin of all the Dhofar buildings. On the site of the 'Oracle' mentioned by Ptolemy were many ruins, including a massive wall of undoubted Sabaean origin. The harbour of Moscha, now nearly blocked by a sandbank, 'is still deep, and extends inland about a mile and a half, and there are many ruins around it. Here we have the *Portus Nobilis* of the 'Periplus,' the harbour to which the frankincense merchants came . . . and where Arab dhows still find a shelter during the north-east monsoons.

Here, I think, the elusive Ophir itself may at last be 'run to earth.' All the factors required for the solution of this perennial problem seem, as it were, brought together in Gen. x. 25–30, where are mentioned in intimate association Joktan (Kahtan), Sheba (Saba), Ophir, Havilah, Mesha (Moscha), and Sephar, in the order here given. Bent tells us that KAHTAN still survives, the local Gara tribe being divided into families, 'the chief of which is the Al Kahtan family, and the head of the Al Kahtan family is recognised as the Sheikh by all the Garas.' He was not aware of the importance of this statement, the significance of which appears from the fact that in the national genealogies the Kahtanides are one of the three great divisions of the Arab race. They are the settled and cultured, as opposed to the wild Bedouin or Nomad element, and still form the bulk of the inhabitants of *Yemen* ('South,' or possibly *Yaman*, 'Happy,' as in *Arabia Felix*), and they claim descent from Kahtan, that is, the Joktan, son of Heber, hence are known as *Arab-al-Aribah*, 'Arabs of the Arabs,' the eldest and purest branch of the race. From Himyar ('Red'), grandson of Kahtan, they took the name of Himyarites, the 'Homeritae' of the Greek writers, by whom the epithet 'red' was extended to the neighbouring coastlands and to all the surrounding waters. The term was applied at first to the whole of the Indian Ocean navigated by the Himyarites, but with increasing knowledge was gradually restricted to the Persian Gulf, the Arabian and Red Seas. For us Himyar is the collective name of all the southern Arabs, and in a narrower sense of the Yemenites, whose two great historic divisions were the Minaeans and Sabaeans. Some (Adites, Themudites, and others) were traditionally cave-dwellers, and we have seen that Bent found many of the Dhofar natives still occupying the spacious caverns of the Gara hills.

Dealing with the section of the 'Periplus' which describes the south coast of Arabia, Tozer remarks that some distance beyond Eudaemon (the present Aden) we come to 'a conspicuous head called Syagros (Cape Fartak), which is here dignified with the title of "the greatest promontory of the world." On its shore there was a depository [depôt] of frankincense, for the neighbouring region of the interior of Arabia (the Hadramaut) has been in all ages the chief source of the supply of that article. Beyond Syagros we meet with another important emporium of frankincense, called Moscha [*cf.* the Meshech of Ezekiel xxvii. 13, which is the Mosoch of the Vulgate], the mountains at the back of which were inhabited by cave-dwellers.'[34]

So also Bent: 'The Gara live chiefly, as stated above, in the deep caves of their limestone mountains, which provide accommodation for the family and many head of cattle. They have a large number of milch cows and goats, and make ghee [clarified butter] in great quantities, which is exported from here. All their implements are of the most primitive description. The churn is a skin hung on three sticks which a woman shakes about until butter is formed ; to make their cows give milk freely they stretch a calf's skin on two sticks, and give this to the cow to lick. The calves and kids are kept in the innermost recesses of the caves during the absence of the dams at the pasturage.'

Such are the primitive ways of these homely troglodytes, as they existed under the shadow of Sephar, probably thousands of years before the neighbouring port of Ophir rose to pre-eminence as the great emporium of the eastern world. For the Semite, left to himself, never changes, and our latest explorers still find them dwelling in their old limestone caves two thousand years after Ophir itself has disappeared, leaving nothing of its ancient glories except the piles of ruins strewn thickly over the plains at the foot of the 'mount of the east.'

CHAPTER VIII

OPHIR THE 'METROPOLIS' AND MOSCHA ITS 'HARBOUR'

OF the thirteen children of Joktan mentioned in the table of genealogies under discussion, we are here concerned only with four—Hazarmaveth (v. 26), Sheba (v. 28), and Ophir and Havilah (v. 29). Of these and all the others, we are told that 'their dwelling was from Mesha as thou goest unto Sephar, a mount of the east' (v. 30). This statement accords well with Sayce's already quoted remark, that the table itself is not ethnological but geographical. From this standpoint its meaning is plain enough. It assigns the whole of South Arabia as far east as Sephar to the Joktanides, or, as they call themselves, the Kahtanides, the 'Arabs of the Arabs,' whom we have collectively called Himyarites, and whose domain from prehistoric times down to the irruption of the Moslem Ishmaelites was precisely this south-western section of the peninsula.

The first name, taking them in their Biblical order, is Hazarmaveth, which is unanimously identified with the Adramitae of Ptolemy, that is, the present Hadhramout, lying, as it ought to do, west of Sephar. Although the word means 'Court (region) of Death,' Hadhramout was not then, any more than now, entirely a waste. Ptolemy even gives it a seaport, the well-known Kane, which lay about midway between Aden and Moscha, and drew sufficient supplies from the fertile tracts (oases) of the interior to feed a flourishing export trade. Amongst these commodities are mentioned frankincense, aloes, and other 'spices for the king,' which were shipped for the lands to the east, with which Kane had an established intercourse.[35] Hadhramout still continues to forward frankincense, but now chiefly by the very old overland routes northwards to Nejd and westwards to Yemen. The centre of the traffic is the inland town of Shibahm, nearly due north from the haven of Mokalla. Here are caravan tracks 'polished by the soft feet of millions of camels that had slowly passed over it for thousands and thousands of years' (Bent, 'Southern Arabia,' p. 89). From this source the Queen of Sheba must have drawn much of her supplies by the western route which still conveys these spices to Sanâ. 'From personal observation it would appear that the ancients held communication with the Hadhramout almost entirely by the land-caravan-route, as there is absolutely no trace of great antiquity to be found along the coast-line, whereas the Wadi Hadhramout itself and its collateral branches are very rich in the remains of the ancient Himyaritic civilisation' (*ib.*).

Evidently the whole region must have been far more productive and more thickly peopled in former times than at present. The highway from Kane to the royal residence, Sabbatha (Shabwa), passed by the ruins of Nakb el-Hajar, which already lay within the frankincense land ruled over by the Sabaean king Eleazos (Il'azz), and has been identified by Glaser (ii. 175) with the present Hiṣn el-Ghuráb five days from Sabbatha. It is important to fix the position, as the ruins of Nakb el-Hajar are brought by Bent into relation with those of Rhodesia. 'Our information of these countries [South Arabia, &c.] is meagre [but greatly extended since these words were written], but some of those buildings which are known in Yemen, which seem to combine temple and fortress in one, as on Zimbabwe Hill, may have been built by the same race that constructed Zimbabwe; and the elliptical temples at Marib and Sirwah, and the one at Nakab el-Hajar, with its north and south doorways seeming to indicate an observation of the meridian, may embody some of the mathematical principles illustrated by the ruins of Mashonaland. When the original builders of Zimbabwe have been traced to their home [South Arabia], it will remain to discover who were their successors in Mashonaland [the Phœnicians] that rebuilt the western wall of the great temple and some portions of other buildings, for this certainly was not done by any of the present negro [Bantu] races.'[36]

After describing Kane, the 'Periplus' (§ 29) writes: 'After Kane, where the mainland recedes farther back, follows another bight reaching far inland and extending a long way, the so-called Sakhalitic and incense-yielding district.' In § 30 the position is determined by the

Syagros headland [see Ritter above] where is a stronghold to protect the country, a harbour, and a frankincense depôt.' Then at § 32: 'After Syagros immediately follows a gulf which stretches far inland; Omana, with a circuit of 600 stadia [furlongs nearly], and beyond it high rocky and steep cliffs where people dwell in caves for a further 500 stadia, and beyond these the harbour called MOSCHA, a wharf known as the emporium for the Sakhalitic frankincense.'

On this Glaser remarks (ii. 178): 'Syagros is clearly the Ras el-Fartak. Here grows the best incense still by the Mahras called *mghar*, and this is from the same root as the *mokroto* mentioned in 'Periplus' § 11 as exported from the port of Mosyllon on the Somali coast. [It grew and still grows on both sides, both in the Arabian and African 'Punt.'] Mokroto seems to be merely a diminutive form like the present [Koranic] Arab *mghairot*, which is the same word. One of the best kinds of incense shrubs is still called mohr=*moghr* in Somaliland [which gives the true source of our word *myrrh*].

'Pliny locates the Haḳḳili on this coast, which extends from the Ras el-Kelb towards Mirbat [see Ritter above], but more particularly comprises the Ṫafâr (Dhafar) district. The Haḳḳili (Fresnel's Ehkili) still dwell here.'

Further (p. 180) Ṫafâr and Abissa Polis are regarded as one and the same place, of which Moscha Portus was the harbour, close to where Ptolemy locates his SAPPHAR METROPOLIS, our OPHIR. Lastly Moscha is identified with Khôr el-Belîd, where the chief ruins are and where the harbour is now partly silted up. 'At present there is no longer any harbour here, but apparently a closed basin, which even in the opinion of the surrounding inhabitants was only later cut off from the sea by a partial silting up.' Here are 'the magnificent ruins also identified by Fresnel with Moscha.'

All this was of course written before Bent's visit, and to complete the picture and fix Ophir beyond the possibility of escape, we must return to the Biblical Ophir, which in Gen. x. follows immediately after Sheba, being here mentioned for the first time in the Bible, and that in unexpected association with the gold and gem-bearing Havilah already introduced, also unexpectedly in Gen. ii. and in Gen. x. 7, strange to say, jointly with Seba and Sheba. The association is of deep import, as showing that the two places were in the popular imagination linked together as by a golden chain, and also implying that the connection was as between *exporter* and *importer*—giver and receiver. Obviously both places are not intended to be taken as producers, and as this rôle is already and in the clearest language assigned to Havilah (see above), while Ophir, although mentioned several times in connection with gold, is *never once* spoken of as itself *a gold-yielding land*, it follows that it is to be regarded, and was regarded, as the receiver, the importer, the mart or emporium whence the precious metal and the precious stones brought from Havilah were distributed from remote times over the ancient world. As in the title of our book, the expression always is 'the gold of (or from)

Ophir,' just as in commercial language we now say the diamonds of Golconda, the senna of Alexandria, the coffee of Mocha, Malaga raisins, Smyrna figs, Sisal hemp, these places being the seaports whence such wares are forwarded, not necessarily the districts where they are grown. Such expressions tend to become stereotyped, and as the Yucatan fibre is no longer shipped at Sisal, although still called 'Sisal hemp,' so 'the gold of Ophir' became in a sense proverbial in Biblical language, apparently long after Ophir had ceased to be a distributor of costly foreign wares. Its true relation to the traffic in such produce is at last beginning to be understood by the commentators, and Mr. H. C. Hart, for instance, remarks in reference to the 'peacock' question: 'Hence it is concluded that Tharshish or Ophir was situated on the coast of Malabar or in Ceylon, or else that costly wares *came from there to some intermediate emporium*.'[37] Here we have the usual confusion between Tharshish and Ophir, as if they were both the same place, some distant over-sea land, occasionally visited by the fleets of Hiram and other Phœnicians, the voyage there and back being supposed by many to have taken three years. But the texts make it perfectly clear that they were two different places, and that the voyage to Tharshish and back did not take three years, but was made every three years, exactly like the Spanish 'silver fleet,' which in colonial times sailed every three years from Acapulco in Mexico *via* Manila to Cadiz. Thus 1 Kings x. 22: 'For the king had at sea a navy of Tharshish with the navy of Hiram: *once in three years* came the navy of Tharshish, bringing gold and silver, ivory, and apes, and peacocks.' The expression 'navy of Tharshish,' twice repeated, would almost imply that Tharshish, like Ophir, was a naval station of the home land. But Jerome's more accurate Vulgate shows that Tharshish was some foreign place to which Solomon's navy sailed once in three years, returning thence with the costly wares.[38]

More important for us is the statement in 1 Kings x. 15 that besides the 666 talents of gold brought to Solomon in one year (should be *every year*, '*per annos singulos*'), he received more gold 'of the merchantmen, and of the traffick of the spice merchants, and of all the kings of Arabia, and of the governors of the country.' The mention of spices and kings of Arabia brings us back to Ophir, which may now with some confidence be located in South Arabia, since a distinction is here drawn between the independent kings (of Yemen) and those parts of North Arabia (Edom, &c.) over which, after their conquest by the king, 'governors' had been appointed. Now in South Arabia, with its forbidding rock-bound coast, there are but few places to choose from, and of these incomparably the most likely is the above-described MESHA or MOSCHA, which in Gen. x. follows immediately after Ophir and Havilah. This juxtaposition, which cannot be accidental, is already significative, but becomes still more so when we read (Gen. x. 30) that 'their dwelling was from Mesha, as thou goest unto Sephar, a mount of the east.' Here it is to be noted that Moscha, another form of which is Mocha, means, like the present *bender* (Bender-Abbas, Bender-Muriyah, Bender-Gori, &c.), a harbour or haven in general, and that of this particular Moscha Arrian tells us that in his time (second century A.D.) it was still much frequented for the Sakhalitic incense obtained there.[39] This fixes the spot because the Sakhalitic bay lay close by, and in fact is the eastern section of Bent's frankincense-yielding Dhofar. We have seen what a splendid harbour this Moscha was in former times, when it was commonly known as the *Portus Nobilis*, or, as one might now say, the Harbour in a pre-eminent sense. I therefore infer by a process of elimination and on the grounds here specified that this Haven *par excellence* was in fact the port of Ophir, which itself stood a little inland, round about the head of the inlet, which Bent tells us is surrounded by 'many ruins,' and was reached 'from Mesha as thou goest unto Sephar.' Mesha, the identity of which with Moscha cannot be doubted, was thus the port of Ophir, which was itself accessible to sea-going vessels by the above described long, deep channel now silted up at its mouth. Ophir was thus the port of entry and the distributor of foreign wares which arrived every three years, and included especially gold, thence called the 'gold of or from Ophir.' It was also the outlet for the local produce, more particularly the frankincense and myrrh which 'the spice merchants' and 'all the kings of Arabia' forwarded to the Court of Solomon, and which drugs are still here shipped in Arab dhows for Bombay to the yearly amount of about 9,000 cwt.

Ophir was further a Sabaean and not a Phœnician port, and we have seen that in this district alone still survives the ancient Himyaritic tongue locally known as Ehkili. On this Ehkili language I contributed a short article to Cassell's 'Storehouse of General Information' (1892), in which it is stated that 'it differs greatly from current Arabic, and represents, in a corrupt form, the primitive Himyaritic speech of Arabia, which survives also in the Tigré and other dialects of Abyssinia on the opposite side of the Red Sea. Ehkili is consequently one of the most primitive dialects of the primordial Semitic language, and in it are composed the numerous rock inscriptions scattered over South Yemen and Hadhramout. From these imperfectly deciphered inscriptions, and from the specimens of Ehkili collected by Fresnel and others, it appears that Himyaritic differs more from Arabic than Arabic does from Hebrew (and Phœnician), with which languages it has much in common.

The importance of this statement will presently be seen, and meanwhile it may be asked, what other seaport in the Arabian waters answers to the conditions here specified at all so well as this Mesha of Genesis, the *Messa* of the Vulgate, perhaps the Meshech of Ezekiel, and Arrian's Moscha or *Portus Nobilis*, which I identify with Ophir? From what can be gleaned from Pliny, Ptolemy, and the 'Periplus,' only three rivals are possible: Muza, Arabia Eudaemon, and Kane. Glaser makes a strenuous effort to locate the Muza of the 'Periplus' (§ 24) in the neighbourhood of Mokha, identifying it with Mashâliha, that is the Masala of the Homeritae [nearly all the MSS. have Nomeritae, not otherwise known], mentioned by Pliny in xii. 69, where he describes the various kinds of myrrh. Amongst these is an excellent white kind, but occurring only in one place, whence it is brought to the city of Mesalum.

So this must be Mesala, which is then removed to the south-west coast of Arabia, that is to Muza, near Mokha; and as Muza, wherever situated, was a famous old emporium, the chief, if not the only mart for myrrh in the time of the 'Periplus,' 'Mashâliha, near Mokha, answers perfectly well' (ii. p. 138). But Pliny and the 'Periplus' tell us that the voyage from Berenike on the Red Sea coast of Egypt takes about thirty days to Ocelis, or to the incense-yielding Kane district. Then Pliny adds that there is a third harbour besides Ocelis and Kane, that is, Muza, which is frequented by the traders trafficking in frankincense and Arabian spices. The station lies inland, where the royal residence is Sapphar. On this Glaser becomes quite unintelligible (p. 139), not knowing that Muza (Moscha) meant a harbour in general. Hence there might be 'a third,' as Pliny says, or, in fact, any number. For us the passage is perfectly clear, and proves convincingly that Pliny's Muza was not Mashâliha, near Mokha, but our Moscha, the harbour of Ophir. Muza thus disappears as a rival, because it is Moscha itself, and this might on reflection be accepted by Glaser, since at p. 168 he admits that his Muza lay inland ('landwärts') from Mokha, as indeed it does to this day. Consequently it is not a harbour at all, nor did it lead to any place in particular, while our Moscha was a magnificent harbour, one and a half mile long, leading to the royal residence of Sapphar, or Ophir, 'as thou goest unto Sephar, a mount of the east.' The road, it may be added, traverses the richest libaniferous land in Arabia.

Dealing with the Mesha and Mount Sephar mentioned together in verse 30 of Gen. x, Glaser remarks (p. 437) that Sephar is a frequently recurring geographical name, with such variants as Safaira, Safâr, Es-Safîr. But he admits that we have here a place either on the south coast of Yemen, or on the frankincense coast, that is, Bent's Dhofar, or in Omân, still farther east; in any case in the south, though he still doubts whether it is the same as his Ṫafâr. Then follows what to me seems a strange *non sequitur*, that is to say, if Mount Sephar is in the south, then Mesha must be many hundreds of miles away to the north. He accordingly suggests that it may be the Jebel Shammar of north central Arabia, which he identifies with the Mash of the Assyrian records, which Mash includes Massa. 'But in that case,' he goes on, 'Massa must be identical with Mesha,' and he asks almost in desperation whether all three—Mash, Mesha, and Massa—may not be arbitrary spellings of one and the same name in the Masoretic texts, or from different documents. He seems himself to feel that the reasoning is unsatisfactory, as the result assuredly is. For two places brought close together in what Sayce tells us is a geographical, not an ethnological, document (see above), are here violently divorced, one being left on the south coast of Arabia, while the other is spirited away to the far north!

Aden (Arabia Eudaemon, whence some supposed Arabia Felix to have been named) and the Kane of Hadhramout need not detain us. Neither of them has a Mount Sephar, nor an adjacent Sakhalitic Gulf, nor any Sakhalitic incense, nor a limestone range with inhabited caves in the background, nor a Syagros headland (Mirbat) in front, nor the splendid haven strewn with magnificent Himyaritic ruins, which fix the position of Ophir beyond reasonable doubt.

We can now fix the very meaning of the word which, as above explained, has the sense neither of 'riches' nor of 'red.' Its identity with Sapphar and all its variants being ascertained, I find that this Sapphar means a 'capital,' a 'metropolis.' The evidence is not overwhelming, but I think sufficient. Speaking (p. 241) of Ptolemy's above-quoted 'Sapphar Metropolis,' Glaser remarks, truly enough, that it is to be sought in the country of the Mahra people, adding: 'Sapphar Metropolis = Ṫafâr Metropolis, is, strictly speaking, a tautology, since Ṫafâr itself means essentially nothing else than "metropolis." Every large and famous city may accordingly be so indicated. In the present instance we have probably to do with Raidân, which in the "Periplus" also is similarly called Tafâr (*Saphar* or *Aphar*).' Here one of the variants is Aphar, which is obviously the same word as Ophir, and Ophir is thus explained by perhaps the first living Himyaritic scholar to mean metropolis. Thus while Moscha is the Portus Nobilis, the seaport in a pre-eminent sense, the neighbouring city at the head of the deep inlet is the metropolis in a pre-eminent sense. The parallelism is so complete that it might give rise to suspicion did it not rest on the authority of Glaser himself. The Raidân, it may be added, here spoken of was at that time one of the great Himyaritic States. In the rock inscriptions Sabâ (Sheba) and Raidân are often mentioned in close association, and Marib shared with Saphar (Ophir) the honour of being a royal residence, like, for instance, Stockholm and Christiania at present. This is the statement of Glaser, who also thinks (p. 180) that Abissa Polis and Ṭafâr are one and the same place, and that their harbour was Moscha Portus [=Portus Nobilis], close to where Ptolemy places his Sapphar Metropolis. One can scarcely help thinking that had Glaser remembered that Moscha meant a harbour in general, as he knew that Sapphar or Aphar meant a capital in general, he would have anticipated the views here advocated, and relieved me from the necessity of writing the present work.

It now becomes all the more easy to understand how Ophir, meaning a capital city *par excellence*, and being for thousands of years the great emporium especially for the gold imported from foreign parts (Havilah), the expression 'gold of Ophir' became, as above shown, proverbial in current speech. Moscha was itself a mere landing or loading place, a wharf (see above); but Ophir was the great city, *the city*, like the *urbs* of the Romans, or as we speak of going *to town* meaning to London.

CHAPTER IX

THE TWELVE OPHIRS—THARSHISH AND HAVILAH

BUT we have now to consider how far all this harmonises with the actual references made in Scripture to Ophir, Tharshish, Havilah, and the incidents mentioned in connection with the gold and other treasures brought to the courts of David and Solomon. I think I shall be able to show that the harmony is complete. Ophir is mentioned altogether twelve times, and it will be convenient here to collect the passages:—

(*a*) Gen. x. 29: And Ophir, and Havilah, and Jobab.

(*b*) 1 Kings ix. 28: And they came to Ophir, and fetched from thence gold, four hundred and twenty talents, and brought it to King Solomon.

(*c*) 1 Kings x. 11: And the navy also of Hiram, that brought gold from Ophir, brought in from Ophir great plenty of almug trees, and precious stones.

(*d*) 1 Kings xxii. 48: Jehoshaphat made ships of Tharshish to go to Ophir for gold: but they went not; for the ships were broken at Eziongeber.

(*e*) 1 Chron. i. 23: And Ophir, and Havilah, and Jobab.

(*f*) 1 Chron. xxix. 4: Even three thousand talents of gold, of the gold of Ophir, and seven thousand talents of refined silver, to overlay the walls of the houses withal.

(*g*) 2 Chron. viii. 18: And Huram sent him [to Solomon], by the hands of his servants, ships, and servants that had knowledge of the sea; and they went with the servants of Solomon to Ophir, and took thence four hundred and fifty talents of gold, and brought them to King Solomon.

(*h*) 2 Chron. ix. 10: And the servants also of Huram, and the servants of Solomon, which brought gold from Ophir, brought algum trees and precious stones.

(*i*) Job xxii. 24: Then shalt thou lay up gold as dust, and the gold of Ophir as stones of the brooks.

(*j*) Job xxviii. 16: It cannot be valued with the gold of Ophir, with the precious onyx, or the sapphire.

(*k*) Ps. xlv. 9: Kings' daughters were among thy honourable women: upon thy right hand did stand the queen in gold of Ophir.

(*l*) Is. xiii. 12: I will make a man more precious than fine gold; even a man than the golden wedge of Ophir.

As Tharshish also occurs in close connection with Ophir, and is often confused with it by the commentators, some holding that they were one and the same place, the pertinent passages may here also be introduced :

(*m*) Gen. x. 4 : And the sons of Javan : Elishah, and Tarshish, Kittim, and Dodanim.

(*n*) 1 Kings x. 22 : For the king had at sea a navy of Tharshish with the navy of Hiram : once in three years came the navy of Tharshish, bringing gold, and silver, ivory, and apes, and peacocks.

(*o*) 1 Kings xxii. 48 as above (*d*).

(*p*) 2 Chron. ix. 21 : For the king's ships went to Tarshish with the servants of Huram : every three years once came the ships of Tarshish bringing gold, and silver, ivory, and apes, and peacocks.

(*q*) 2 Chron. xx. 36, 37 : And he [Jehoshaphat, king of Judah] joined himself with him [Ahaziah, king of Israel] to make ships to go to Tarshish ; and they made the ships in Ezion-gaber. Then Eliezer, the son of Dodavah of Mareshah, prophesied against Jehoshaphat, saying, Because thou hast joined thyself with Ahaziah, the LORD hath broken thy works. And the ships were broken, that they were not able to go to Tarshish.

(*r*) Ps. lxxii. 10 : The kings of Tarshish and of the isles shall bring presents : the kings of Sheba and Seba shall offer gifts.

(*s*) Ezek. xxvii. 12 : Tarshish was thy merchant by reason of the multitude of all kind of riches ; with silver, iron, tin, and lead, they traded in thy fairs. And verse 25. The ships of Tarshish did sing of thee in thy market: and thou wast replenished, and made very glorious in the midst of the seas.

Lastly, one passage must be added, where reference is made to the 'gold of Parvaim.'

(*t*) 2 Chron. iii. 6: And he [Solomon] garnished the house with precious stones for beauty: and the gold was gold of Parvaim.

Here we have all the passages in any way bearing on the subject under discussion. The bracketed letters by which these are indicated will now serve for reference purposes, whereby clearness and brevity will be consulted.

As Parvaim occurs once only, it may be summarily disposed of. If it is not the real name of Ophir, which, as explained above meant only metropolis, the expression 'gold of Parvaim' must still have had much the same proverbial sense as 'gold of Ophir.' It is introduced apparently before the alliance had been contracted between Solomon and Hiram, and as Glaser aptly remarks (p. 347), 'this gives one the impression as if Parvaim gold and Ophir gold were one and the same; for it is highly improbable that the same Hiram would now suddenly have sought out another gold land besides that hitherto visited. Hence we must seek Parvaim in Ophir, or better in Havilah.' With this I am in full accord, but not with Glaser's identification of Parvaim, the *Sâk el-Farwain* of North Arabia, which is not even described as a mining district.[40] In other respects Parvaim interferes in no way with our general thesis.

Touching Tarshish or Tharshish we know that there

were several places so named, while others have been included which have no claim to the distinction. Such is the Tartessus of Iberia (Spain), which is commonly but wrongly identified with the Tarshish of (*d*), (*n*), &c. Sayce amongst others writes (*op. cit.* p. 47): 'Tarshish is usually identified with Tartessus in Spain. It was the farthest point reached in the western basin of the Mediterranean by the Phœnician and Greek traders. The ships which made the voyage were consequently known as the ships which traded to Tarshish, or more briefly, 'ships of Tarshish.' The phrase gradually came to be applied to any kind of merchant vessel, even to those which had never visited Tarshish at all.

So also Tozer much to the same effect (*op. cit.* p. 7): 'The neighbouring region of southern Spain was known through them [the Phœnicians or their Carthaginian descendants] as Tarshish or Tartessus—a name which was derived from the tribe that inhabited it, the Turti or Turdetani. Of this we hear even in the genealogies in the tenth chapter of the book of Genesis [*m*], and in Solomon's time (1000 B.C.) it is mentioned in connexion with the navy of Hiram, king of Tyre' [*n*].

Here is a curious blend of truth and what must be called fiction, which was partly due to popular etymologies, but in any case has a venerable antiquity, dating certainly from the time of the Alexandrian Greeks, and possibly even from that of Ezekiel himself. The Tarshish of (*m*) was equated with the Iberian *Turti* or *Turdetani*, although they were of Hamitic stock, like the Basques still surviving in north-west Spain,[41] whereas the Tarshish of (*m*) was a 'son of Javan,' that is an

Asiatic Greek, an Aryan (Javan = Ionian), and this Tarshish was in Cilicia,[42] not in Spain, at the other end of the Mediterranean, or rather on the Atlantic Ocean, a little beyond the Pillars of Hercules.

But was Tartessus a seaport at all? Was it not rather the name of the deltaic island at the mouth of the Baetis (Guadalquivir), where Strabo, Pausanias, and others locate the famous seaport of Gades (*Portus Gaditanus*), the present Cadiz? But allowing that it was a town or some small place close to, though distinct from, Gades, historical geographers will still remember that inland from the Turdetani were the kindred Turduli, whose capital was *Illiberis* (*Eliberris*), a word which also occurs elsewhere in Iberia, and even beyond it, and is still good Basque for *Newtown*. 'When Hannibal crossed into Gallia Narbonensis on his march to Italy, he came upon a flourishing city *Illiberis*, a name with which his Iberian allies were familiar, because they had left behind them in their own territory of Baetica (Andalusia) another place of the same name meaning in their language "Newtown," as it still does in modern Basque.'[43] All these ancient Andalusians were therefore Iberian Hamites from North Africa, and not Hellenic sons of Javan from the Eurasian steppe lands. But the names seemed near enough, and so the Tarshish merchants became 'Carthaginians' in the Septuagint version of (*s*), because they had settlements in Spain, such as Carthagena, and especially Gades, close to the Tartessus of the Turdetani.

But it was forgotten that Spain could not supply all that 'multitude of all kind of riches' referred to in (*s*), the ivory, for instance, and the peacocks expressly mentioned in (*n*). For these commodities, even if the 'peacocks' were only 'guinea fowl,' we must go to the Indian Ocean, where some other Tarshish—that of (*d*), (*n*) and (*p*)—must have existed. But the difficulty is to get there before the opening of the Suez Canal. There was certainly a natural waterway between the Mediterranean and Red Seas in Pleistocene times, say, 100,000 years ago, and although that was gradually closed by upheaval, it is said to have been indirectly restored by the Pharaohs of the nineteenth dynasty some 3,400 years ago. A tradition mentioned by Strabo attributes the restoration to Sesostris (Ramses II.), to whom so many exploits were credited. It appears to have really been the work of his father Sethi (Meneptah I. 1366 B.C.) ; it did not however run straight from sea to sea, but from a branch of the Nile delta at Bubastes, near the present Zagazig, eastwards to some point above the head of the Gulf of Suez. This canal appears to have been again closed, perhaps by the blown sands, and Herodotus tells us that another, running much in the same direction, was begun, but never finished, by Neko (Necho), son of Psammeticus (620 B.C.). This is the king who sent a Phœnician fleet to circumnavigate Africa, and there is good reason to believe that the expedition was successful. Not so the canal, which, after costing the lives of 120,000 hands, was abandoned because of an oracle declaring that it would be utilised by foreigners. It was so utilised by Darius of Persia, who opened a wide communication between the Nile and the Gulf of Arsinoe, and even began another running directly to the same point from Pelusium, on the Mediterranean, that is, along the course of the present Canal. The attempts were renewed under the Ptolemies, and again more successfully under Trajan, and we are told by Makrizi that the 'river' named from that emperor was still navigable between the Nile and the Bitter Lakes for some time after the Moslem conquest of Egypt. It was even cleared and improved by Amru, although the restored canal did not last very long, 'having been finally closed 130 years afterwards by order of the Caliph Abu Jafar el-Mansur, to prevent some rebel from receiving his supplies.' [44]

Here we are concerned only with the Pharaonic communication, which, although presenting a roundabout way between the two seas, might very well have been used by the Phœnicians, who, we know, had settlements in Lower Egypt, and might have found it profitable to pay toll to the Egyptian kings for the right of way. On this subject Mr. Andrew Syme has an instructive letter in 'Rhodesia,' October 5, 1901, in which he remarks that 'the story of this navigable channel is of great importance when dealing with the history of the Phœnician and Himyarite navigators of South Arabia. According to all traditions and probabilities the Phœnicians were a branch of the ancient seafarers of the Indian Ocean, who had passed through the Straits of Suez and had formed settlements on the Syrian coast, where there was good timber for shipbuilding, as well as at important Egyptian coast and river ports. Memphis, Heliopolis, and Bubastis all lay on the main channel of navigation connecting the Red Sea, Mediterranean, and Nile Valley. Memphis had its Phœnician quarter, and its patron deity had a Phœnician name—Ptah (The Opener). Heliopolis was the "City of the Sun" and of the sacred Red Bird of the South—the Phœnix. The Nile then flowed close by this famous city. Bubastis, the "City of Bast," the cat-headed goddess, to whose peculiar cult the modern world owes the domestication of "puss" (Biss, Bess, are the South Arabian names for cat), was situated at the junction of the Red Sea Channel and the most easterly branch of the Nile to the Mediterranean at Pelusium. It was a place of considerable importance, and must have been frequented by ships and seafarers from many different countries. The Proto-Phœnician or Himyarite navigators of the Greater Red Sea had probably learnt the art of ocean-sailing from that still more primitive seafaring people the Proto-Malays, sometimes called the Oceanic or the Malay-Polynesian race.'

The Malayo-Polynesians have no doubt always been daring seafarers, but we shall see that the Himyarites and Phœnicians did not need to learn the art of navigation from them. Nor do the Jews and Phœnicians appear to have made much use of the Egyptian waterway, since we are expressly told at (*q*) that Jehoshaphat, king of Judah, was already building ships for himself in Ezion-geber, though, to be sure, they were wrecked before sailing for Tarshish. Besides, in the time of Sethi the Egyptians had their heel on the necks of the peoples of Canaan (Palestine, Philistia, Phœnicia). The great stele of that king found by Professor Flinders Petrie in 1896 at Thebes has an inscription celebrating his victories, in which Israel is mentioned by name for the first time. One passage is thus translated by Dr. Spiegelberg: 'Kheta [the Hittites] is quieted—The People of Israel is laid waste—Seized is Canaan with every evil—Led away is Askalon—Taken is Gezer [Gaza?].' It is impossible to suppose that the Phœnicians could have been using a waterway running through Egypt at this period, or even for some time afterwards. About 300 years later they began to breathe again, and it was then that the kings of Tyre and Israel struck up an alliance, warned by previous disasters caused by disunion.

It is to be carefully noted that the passage in (*d*) 'Jehoshaphat made ships of Tharshish to go to Ophir for gold,' is explained at (*q*), where Tarshish takes the place of Ophir, and is repeated twice: 'to make ships to go to Tarshish;' and 'And the ships were broken that they were not able to go to Tarshish.' It is precisely the same in the Hebrew and Vulgate, so that Tharshish is obviously to be distinguished from Ophir as the *terminus ad quem*, the objective, while Ezion-geber is the *terminus a quo*, the port of clearance at the head of the Red Sea, Ophir, as we shall see, lying about midway between the two.

Our Tharshish is therefore some seaport in the Indian Ocean, and not in the Mediterranean, or on the Iberian shores of the Atlantic, where no ivory or peacocks were to be had. And if it be asked how a seaport of this name strayed from the northern into the southern waters, the explanation seems simple. The Phœnicians were familiar with all of the Mediterranean Tharshishes, some, or even all, of which they may themselves have established, for the word is undoubtedly Semitic, while several were stations in or near mining districts—the Cilician Tarsus, for instance, near Kittim, that is, Cyprus, noted for its copper mines.[45] Thus they brought the name with them into the Indian Ocean, where they founded another Tharshish as the outlet for the gold of Havilah (Rhodesia), by far the richest auriferous region known to the ancients. So also the modern British settlers in Austral Africa and Australia have now their 'Newcastles' in Natal and New South Wales, just as the Iberian Basques had their 'Newtowns' in Spain and Gaul.

But Tarshish is not found in the Indian Ocean, it will be objected. The name not, but the thing yes, and names are but symbols of things. And if not the thing, then one may ask, what Tarshish was that to which the fleets of Solomon and Hiram, outward bound from Ezion-geber, went in quest of the gold and ivory which they could find neither in the Mediterranean nor in the Atlantic, nor in India, nor yet in Arabia, nor in Ophir itself, now located on the non-auriferous Himyaritic seaboard? Wherever they found the gold and the ivory, there also was their Tarshish, and that was in the ancient Rhodesia, the Havilah 'where there is gold,' for there is no other. And if both names, Havilah and Tarshish, have disappeared, they have but shared this fate with hundreds of others, effaced, washed away past recovery by the ebb and flow of streams of racial movements continued over thousands of years. And again, if the names are gone, the stones remain, and Shakespeare tells us there are 'sermons in stones.' The sermon that the Rhodesian stones preach with convincing eloquence has already been clearly indicated, and will here again be enforced.

I may remark in passing that Havilah, if an ethnical term at all, was of Himyaritic origin like Hazarmaveth, both being included amongst the Children of Joktan (Gen. x. 26 and 29), while Tarshish was almost certainly Phœnician. But since the withdrawal of the Himyarites and Phœnicians, Rhodesia and its seaboard have been swept by successive floods of Moslem Arab and Bantu invaders, many Bantus, such as the Zulu-Xosas, the Bechuanas, the Hereros, all setting from beyond the Zambesi to their present southern homes. As well therefore go in search of the terms Tarshish and Havilah as of Almodad, Sheleph, Jerah, and all the other 'lost tribes' of Joktan. Hazarmaveth has been exceptionally preserved in the lottery of nations, and its persistence is really wonderful, for the present Hadhramout answers letter for letter to its Himyaritic precursor, and to the Hebrew Hazarmaveth. 'There is every reason to believe that anciently, too, the Hadhramout meant only this valley; we learn from Himyaritic inscriptions that five centuries B.C. the name

t m r d h

was spelt by the Himyars as it is now (namely XΣ)ΒΨ [to be read from right to left]), and meant in that tongue "the enclosure or valley of death," a name which in Hebrew form corresponds exactly to that of Hazarmaveth of the tenth chapter of Genesis.' [46]

It might here, perhaps, be argued that Rhodesia could not be Havilah, because Havilah is in Arabia. The name occurs several times, and in some passages in such a way as to imply that there was certainly a Havilah in Arabia. Thus Gen. xxv. 18: 'And they dwelt from Havilah unto Shur, that is before Egypt, as thou goest towards Assyria,' and elsewhere. But here was no gold, indeed mostly nothing but sands, and the sacred texts clearly imply that our Havilah, the auriferous Havilah, was not in Arabia, since it could not be said of this region that 'the gold of that land is good; there is bdellium and the onyx stone.' These, as we have seen, are products of Rhodesia, although there might be a doubt about the 'onyx stone.' In the Hebrew the word is *shoham*, and Dr. Pinches writes me (August 26th, 1901) that 'if *shoham* be the Babylonian *sâmu*, as is very probable, I should prefer "chalcedony" to "onyx."' Now chalcedony, mentioned also in Revelation xxi. 19, is a cryptocrystalline variety of quartz, a true quartz with some disseminated opal-quartz, hence is appropriately associated with bdellium, the krustallos of the Septuagint. It will, therefore, most probably be found amongst the various allied crystals abounding in some parts of Rhodesia (Matabililand).

On the other hand, the remarkable association of Havilah with Ophir, which was in Arabia, has been above explained by the 'golden link,' the close commercial relations already established between these two places at the very dawn of history.

Coming now to the above specified twelve Ophirs, the first (*a*), with which must be included (*e*), is most interesting, but also perplexing. Havilah had already been unexpectedly introduced into Gen. ii. 11, as the gold land in a superlative sense. How comes it to be here, one should say also unexpectedly, again introduced now as a son of Joktan, a brother of Ophir? Of course, as already shown on the authority of Sayce, these are not really racial but geographical names. Now when Genesis was written—nobody is any longer bound to believe that it was written by Moses, for even the strictest orthodox expounders give this up—the traffic in gold was flourishing. A veritable 'Pactolus' was streaming into Tyre and Jerusalem. The source of this stream was in Havilah 'where there is gold; and the gold of that land is good.' Notice the word 'land,' and bear in mind that Ophir was above shown to be a famous metropolis, to whose Moscha ('harbour') the gold and the other costly wares were consigned, and there re-shipped for Tyre and Palestine. In the popular imagination Havilah and Ophir were thus intimately connected by this golden link, and when Ophir, a real place, a well-known metropolis, *the* 'metropolis,' was mentioned, it was almost inevitable that Havilah should also be mentioned in association with it. One was the source, the gold *land* itself, the other was the receiver, the distributor, and the two were as one in the mind of the people, on the tongue of everybody, and so Havilah, the 'land,' was in uncritical times transformed with Ophir to a child of Joktan. Lately I have heard people, more or less intelligent people, speak in the same breath of San Francisco and Klondike, in such a way as to imply that in their mind the two were one, or all events not far apart. Yet they are about as distant from each other as was Ophir in South Arabia from Havilah beyond the Zambesi.

Several of the other Ophirs—(*b*), (*c*), (*g*), (*h*), some of which are indeed mere repetitions in Chronicles of passages in Kings—now become perfectly clear. All tell the same story of the alliance between Hiram and Solomon, whose fleets unite at Ezion-geber, whence they clear for Ophir, and here ship the gold and ivory and spices and all the other rich merchandise, some of which was brought thither from the auriferous land of Havilah, some likely enough from India also, and some, especially the frankincense and myrrh, procured from the neighbouring district on the slopes and at the foot of Sephar, 'a mount of the east'—east from the west coast of Arabia, where lay Ezion-geber. When rightly understood, this tenth chapter of Genesis is wonderfully accurate as a geographical record, and the reason is because the whole of the Arabian seaboard was at this time very well known to the Jews through their Phœnician allies and the highly civilised southern Himyarites.

It has been shown that Rhodesia possessed all the commodities supplied to Solomon, except, perhaps, the 'peacocks,' if such they were, and these may now be safely presented to those philologists who still believe that Ophir was India, but have nothing to say about Havilah, which was Rhodesia. Anyhow, the peacocks could very well have come to Ophir from Taprobane (Ceylon), and from India, with which regions the Sabaeans also traded. I have called Tharshish the outlet of Havilah, which would identify it with Sofala or Beira, or any other convenient station on the 'Costa de Caffres' of the early maps. And if it be asked, why could not the gold and the silver also come from some Indian Tharshish, the answer has already been given—that, in that region there is very little gold or silver, and certainly no ancient workings which, like those of Rhodesia, could have supplied David with his 'three thousand talents of gold, of the gold of Ophir, and seven thousand talents of refined silver' (*f*), and Solomon with his yearly revenue of 'six hundred threescore and six talents of gold' (1 Kings x. 14).

The (*f*) here quoted has an important bearing on the question, because it shows that the traffic in the precious metals was prior to the time of Solomon, and was already flourishing during the reign of his father, David. It is David who is here speaking, and in verses 2 and 3 of the same chapter xxix. of 1 Chron. he 'said unto all the congregation . . . Now I have prepared with all my might for the house of my God the gold for things to be made of gold, and the silver for things of silver. . . . Moreover, because I have set my affection to the house of my God, I have of mine own proper good, of gold and silver, which I have given to the house of my God, over and above all that I have prepared for the holy house.' Then verse 5: 'The gold for things of gold, and the silver for things of silver.' The stress here laid on the gold and silver implies that the precious metals had already been stored up in abundance for the future building of the Temple by Solomon, so that the traffic must at that time have already been of long standing.

Of our twelve Ophirs there now remains only (*d*), which has been discussed farther back, where I deal with the waterway between the Mediterranean and the Red Sea; and lastly (*i*), (*j*), (*k*), and (*l*), which do not need to be discussed. They occur in later times, and belong to the category of those passages where the 'gold of Ophir,' like the 'diamonds of Golconda,' fades into the region of the fabulous—*nominis umbra et praeterea nihil.* This suggests another explanation of (*t*). Parvaim may very well have been some ancient goldfield—perhaps in north-west Arabia, or in Karamania, east of the Persian Gulf—which became exhausted, or else eclipsed by the fame of Havilah. The expression 'gold of Parvaim' may then have still lingered on for a little while, till at last superseded by the 'gold of Ophir.'

CHAPTER X

THE HIMYARITES AND THEIR MONUMENTS—RHODESIAN PARALLELISMS

BUT *majora canamus!* We pass now from their surroundings to the Himyarites themselves. The term HIMYAR we have taken (see above) as a convenient collective name of all the JOKTANIDES (KAHTANIDES), that is, that great southern and settled section of the Arab race which has been sharply distinguished throughout historic times from the northern nomadic section, commonly and often by themselves called ISHMAELITES. Glaser has a very instructive passage on the land of the Joktanides and the land of the Ishmaelites, which is now (since about 300 B.C.) all called Arabia, but was not so in early times. 'The inscriptions of the Assyrians and Achaemenides [Persians], as also the Minaeo-Sabaean inscriptions, and no less all the Biblical passages where *Arabia* and the *Arabs* occur, show clearly and unequivocally that by the land of the Arabs (mât עֲרָב, עֲרַב) was not then understood what we now call Arabia, but the northern section of the peninsula about west and east of Jauf, and a part of the wilderness stretching north of this line. Not till much later, scarcely 300 years before the new era, was the name "Arabia" extended to the whole land. Even at this time no Arabia was yet known to the South Arabian inscriptions' (ii. 315).

Hence the extreme importance for us of distinguishing between the northern and southern sections of the peninsula and their inhabitants, between the Arabia of the Kahtanides and the Arabia of the Ishmaelites. But as both alike are SEMITES in stock and speech, the difference is not easily grasped by the ordinary student who is not a specialist in philology and ethnology. It seems therefore desirable here to explain who are the Semites themselves. We need not go back to that remote epoch when there were neither Semites nor Hamites, but all formed a not yet dismembered Semito-Hamitic family, whose primeval domain was most probably North Africa, between the Mediterranean and Sudan (*Beled es-Sudân*, 'Negroland'). For the present purpose it will suffice to state that from the remotest times, certainly from the New Stone Age, the south-west corner of Asia—the Arabian Peninsula, Canaan, Syria, and the region stretching thence eastwards to Mesopotamia, and merging northwards in the Armenian and Anatolian uplands—has been the home, I think we may say the cradleland, of the specialised Semitic group. I do not know of any other race by which this region was occupied in pre-Semitic times, and it is my belief that it had not yet been inhabited by man when the ancestors of the Semites migrated thither after separating from the western (African) branch of the family.

The expression 'Semitic' is, of course, merely conventional. We do not even know that there ever was an eponymous hero, Shem, from whom the race may have sprung. But we do know that the word was first introduced by Eichhorn, and that not in an ethnical, but only in a linguistic, sense. This eminent philologist really wished to group together all the peoples who spoke dialects of the same mother-tongue, and were assumed to be descendants of Shem. Such were Eber, father presumably of the Hebrews, and also of Joktan, whence our Joktanides (South Arabians); then likewise of Asshur, whence the Assyrians; and of Aram, whence the Aramaeans, better known as Syrians. But the scheme did not and could not answer, because the Biblical tables are not racial, but geographical. No doubt the above-mentioned all spoke Semitic tongues, but so must have others also, such as the Amorites and Sidonites (Phœnicians), who we know spoke a dialect closely related to Hebrew, although they sprang through Canaan from Ham, and were consequently not Semites at all, but Hamites, according to the tables. But the collective term Semite, being convenient, held its ground, and was even extended to all races now or formerly speaking Semitic tongues, because in the infancy of the two sciences of philology and ethnology, everybody was freely arguing from language to race, and contrariwise. Thus was constituted the Semitic family, the main branches of which may thus be tabulated:

JOKTANIDES, KAHTANIDES, HIMYARITES, OR YEMENITES.—Of South Arabia, dominant also from prehistoric times in Abyssinia, include the Minaeans, Sabaeans, Mahra and Sokotra peoples; the Tigré, Amhara, Shoa, and other nations of Abyssinia. Speech, the most archaic and generally best preserved of all the Semitic tongues; is the language of the rock inscriptions.

ISHMAELITES.—The Arabs, commonly so called, of North Arabia; speech also very old, ranking in this respect next to Himyaritic, of the rock inscriptions; is the dialect of the Koreish tribe, of which Muhammad was a member, and has been widely spread with the diffusion of Islám, from Morocco to Malaysia.

ASSHURITES.—The Assyrians, early in Babylonia, later ranged up the Tigris northwards to and beyond Nineveh; speech, the Semitic of the cuneiform inscriptions, younger than Himyaritic and Koranic Arabic; long extinct.

ARAMAEANS.—Of Syria, parts of Palestine, Babylonia, Armenia, North-West Persia, and Asia Minor; speech, intermediate between Assyrian and Canaanitic; two branches: Syrian in the west, quite extinct, and Chaldean in the east, still current amongst a few so-called 'Nestorians' (Nazarenes), in Kurdistan and about Lake Urmia.

CANAANITES.—The Israelites (Hebrews, Jews) of Palestine, the Moabites, Amorites, Philistines, Samaritans, Phœnicians, Carthaginians; speech, even in its oldest known forms, relatively recent; often shows striking analogies with Himyaritic, due, perhaps, to long-standing intercourse between the Jews and Sabaeans; everywhere extinct, but preserved in Holy Writ and on some Phœnician monuments; also a few corrupt lines in Plautus.

Semitic is, perhaps, the most persistent of all forms of speech. The Assyrian of the cuneiform documents differs, for instance, even less from the modern Arabic of Egypt or Syria, than does English from any of the sister Teutonic tongues. The vowels change freely, but the consonants scarcely at all. They shift about according as they are 'moved' by the vowels, but never alter their relative position to each other. The great feature of the system is its so-called 'triliteralism;' that is to say, every root consists in theory of three consonants, which undergo endless modifications of form, without ever disappearing or dropping out of their proper place. *Qatala*, he slew, may become *qâtil*, a slayer, *qutâl*, slain, *qitl* or *qutl*, slaughter, and so on; but *q*, *t*, and *l* must always follow in this sequence to the end of time.

Thanks to this astonishing persistence, we are able to say that the proto-Semites were a pastoral people, dwelling in tents, and not husbandmen living in houses and towns. Thus Sayce points out that the Assyro-Babylonian *âlu* = city, originally meant a *tent* (Hebrew, *ohêl*); *to go home* is literally to go to one's tent, while the Assyrian *êkallu*, and Hebrew *êkâl*, *palace*, is a loan word from the non-Semitic Akkadian *ê-gal*, 'great house.' This almost inexplicable vitality of the Semitic linguistic system will receive further illustration when we come to deal with the Himyaritic elements in the Malagasy language of Madagascar, in connection with the southern movements of the Sabaeans and Phœnicians towards Rhodesia.

It will now be easy to understand what is meant by Semites, and in what relation the various branches of the family stood to each other. We can also understand how readily alliances were contracted between the Jews and Phœnicians, who both belonged to the same sub-group, as well as between these two and the southern Himyarites, all being of the same stock, say, cousins once removed.

Now, resuming our study of the tables, it will at once be noticed that there is a superabundance of Sebas and Shebas (there is another at Gen. xxv. 3), which would be puzzling were Genesis x. to be accepted as an ethnological document. At v. 7 we have both forms, while Sheba recurs at v. 28, the former being, therefore, apparently Hamites, the latter Semites. But this need no longer trouble us, and it will be more profitable to remember that there were always two forms, written in Hebrew with two quite different initial letters,[47] and maintained throughout long periods of time. Thus the two associated together at v. 7 above reappear also in close connection in Ps. lxxii. 10, where also, strange to say, Tarshish is mentioned, while at v. 15 we have actually the 'gold of Sheba.' Thus 10: The kings of Tarshish and of the isles shall bring presents; the kings of Sheba and Seba shall offer gifts. And v. 15: And he shall live, and to him shall be given of the gold of Sheba. The psalm is assigned, perhaps rightly, to David, and the general reference is then obviously to his son, Solomon. Consequently we have both forms, lasting from Genesis to the time of the Kings. This is but in accordance with the above explained persistent character of the Semitic languages, and finds a striking analogy in some Arab tribes, such as the Shahrâns and Sahrâns, the Shaibâns and Saibâns, the Shuraihs and Suraihs, who have for untold ages dwelt in close proximity, while still maintaining their tribal independence. The reader will also recall the case of the hapless Ephraimites, who said *sibboleth* for *shibboleth*, whence

> So many died,
> Without reprieve adjudg'd to death
> For want of well pronouncing shibboleth.
>
> *Samson Agonistes.*

It is immaterial to our case which of the Tarshishes is meant at v. 10, though 'the isles,' that is, 'the Isles of Greece,' would point to Cilicia. Regarding the other terms, Sayce (*op. cit.* p. 65) remarks that 'Sheba is the Saba of the native [rock] inscriptions, whose ancient capital is now represented by the ruins of Mareb, in the south-western corner of Arabia.' Glaser also, even a greater authority on this point, tells us (ii. 399) that wherever Sheba is mentioned in the Bible 'one and the same Sabaean people are meant, as we know from the [rock] inscriptions.' And again (p. 403): 'We learn from the Assyrian inscriptions, coming down to the time of Sargon [the Third], that at that time there was only one great Sabaean land, extending far to the north.' This suffices. Seba may now be relegated, with many commentators, to East Africa, or to Central Arabia, where it may have been swallowed up in the sands of the desert, while Sheba is identified beyond all cavil with the Sheba of Solomon, the Saba of the rock inscriptions, the land of the Kahtanides and Himyarites. Thus also 'the gold of Sheba' becomes 'the gold of Ophir,' that is, of the Sabaean emporium, here located at Dhofar, the frankincense land, and importer of this gold from abroad.

Sheba is therefore for us the land of the Sabaeans—Yemen, Arabia Felix—the 'Queen of the South,' the richest of all known lands, as it is described by the early writers. It extended from the Red Sea eastwards to Dhofar,[48] where Bent came upon the Mahri (Mahra) people, who alone of all modern Arabs still speak a Sabaean dialect (Ehkili). The 'Queen of Sheba,' who has been unnecessarily sought for in Abyssinia, Rhodesia, India, and elsewhere, came not from over the seas, but from Yemen by the historical caravan route, as is obvious from the wording of the text: 'She came to Jerusalem with a very great train, with camels that bare spices, and very much gold and precious stones' (1 Kings x. 2). The camels, which in any case could scarcely have been transported in the light, seaworthy craft of those times, were characteristic of Arabia, unknown even in Egypt apparently till the Ptolemaic period. No reference occurs to this animal on any of the Nile monuments, or in any extant documents much before the New Era. This princess, Balkis, as the Arabs call her, came from Saba of Yemen, and it may be added that in the Assyrian records reference is made to other 'queens of Arabia,' where matriarchal rites are known to have prevailed in prehistoric and later times. Mention is even made of her capital, the renowned city of *Maraiaba Bahramalakum* ('Marib of the Royal Lake'),[49] where she is traditionally said to have constructed the mighty barrier between two mountains, which captured the running waters, and transformed the upland valley into a vast inland sea, a 'royal lake,' comparable to the Egyptian royal Lake Moeris. It was the bursting (160 A.D.?) of this barrage, the greatest calamity recorded in Arabian history, that shook the Sabaean empire to its foundations, and caused a general dislocation of the tribes throughout the peninsula. It prepared the way for the Axumite conquest of Yemen (523-30 A.D.) and explains the feeble resistance offered by the Himyarites to the second deluge of fiery Moslem fanatics, by which the ancient civilisation of the Kahtanides was submerged. This second deluge was 'an explosion that has been quaintly compared to that of the rarely flowering aloe, which vegetates silently for many generations, and then suddenly bursts into a glorious bloom, dazzling the sight with its brilliant effulgence.'[50]

Little now remains to recall the ancient Sabaean empire except the thousands of rock inscriptions, of temples, palaces, and other ruins scattered over the whole of Yemen as far east as Dhofar. Amongst these ruins special mention is claimed by two in the Maraiaba district—the building traditionally known as the 'Palace of Balkis,' Queen of Sheba, and the great temple, the plan of which was found by Bent to present close analogies with that of Zimbabye. 'What appeared at first sight to be a true circle eventually proved elliptical —a form of temple found at Marib, the ancient Saba[51] and capital of the Sabaean kingdom in Arabia, and at the castle of Nakab al Hajar, also in that country' ('Ruined Cities,' p. 93).

From the orientation of these structures it is evident that they were planned with a view to facilitate astronomic observations. In this also the analogy with Zimbabye is very striking. It is remarkable that only stars of the northern hemisphere seem to have been observed at Zimbabye, for in the great temple itself the culminations of southern stars could quite as easily have been observed as those of northern ones, and in the fortress all view of the northerly sky is almost completely shut off by the cliffs and huge boulders which form its northern line of defence; yet every point from which northern stars could have been observed has been used for this purpose, and there is no temple there from which northern stars were not observed, while at the same time the openly displayed southern sky has been left unregarded. 'This of course points to a northern origin for the people, and suggests that before they came to Zimbabye they had acquired the habit of observing certain stars—a habit so strong that it led them to disregard the use of the southern constellations, though they must have known that they would equally well have served to regulate their calendar; it even seems to indicate that they attached ideas of veneration to certain stars, and rendered them worship.'[52] A perfectly analogous case is that of the Central American Aztecs and Mayas, whose primeval homes were in the north, and who at the time of the Discovery were found still regulating their calendric systems and Tonalamatls (astrological records) by the northern constellations, and if Mrs. Zelia Nuttall can be trusted, especially by the revolutions of the Great Bear and North Polar Star.[53] On the Sabaean star-worshippers this diligent student of early cultures writes (p. 322): 'In remote antiquity star-worship prevailed throughout Arabia, and one of its great centres was the flourishing land of Saba or Sheba, whose queen visited Solomon at Jerusalem. The star-cult of the Sabaeans is acknowledged to have resembled that of the ancient inhabitants of Syria, Mesopotamia, Persia and India [and she might have added, had she known it, the Himyaritic settlers in Rhodesia]. . . . It is unnecessary to point out the significant association of an annual count of days with the stable centre, and its importance as an indication that *the ancient Arabian star-gazers originally associated the year period with circumpolar rotation*' (p. 324).

A conspicuous feature—indeed, one might almost say the essential feature—of the temples at Marib and other parts of the Sabaean and the Phœnician lands were the cones or round towers such as that which is seen so plainly on a well-preserved coin of Byblos, a very ancient Phœnician foundation on the coast, north of Sidon. So it was also at Great Zimbabye, where Mr. Swan tells us (p. 127) that 'the most important feature of the interior of the temple is, of course, the great tower, which is a marvel of workmanship in rough material, and in the truth of its lines almost as wonderful as the column of a Greek temple. We could at first discover no reason for its being built in its peculiar position. It has not been placed with any reference to the points of the compass, nor to the bearing of the sun at the equinoxes, and its position is only indirectly connected with the position of the sun at the solstices. But it is in the middle of the space marked off by the two inner doorways, and the more easterly of these two doorways is at the point where the sun would appear when rising at the summer solstice when regarded from the central altar ; and the other doorway is at the point where the decoration on the outer wall terminates, and that is the part of the wall where the sun's rays would be tangential to its curve when rising at the same solstice . . . The towers when built were doubtless made complete in their mathematical form, and were carried up to a point, as we see in a coin of Byblos, where we have a similar tower represented with curved outlines.'

The ruined temples at Marib and in other parts of the Himyaritic lands have not yet been thoroughly explored. But when this work is done, and especially when they are studied in connection with the now well-known Rhodesian monuments, sufficient data have already been collected to make it almost absolutely certain that all alike were built by Semitic star and phallus worshippers. Near an altar in one of the smaller Rhodesian shrines were picked up several phalli, and the Sabaeo-Phœnician towers are, I believe, universally admitted to have been directly associated with phallus-worship.

The same conclusion, which is becoming more and

more inevitable with every fresh turn of the spade, is pointed at by many other Rhodesian finds. Thus the soapstone birds on pedestals, fully illustrated by Bent, are embodiments of a symbolism, which need not here be described in detail, but was undoubtedly common to the ancient Assyrian, Phœnician, and other Semitic peoples. 'In the difficult question of early Arabian cult, which was closely bound up with that of Egypt, Assyria, and Phœnicia, we find the vulture as the *totem* of a *Southern Arabian* tribe at the time of the Himyaritic supremacy, and it was worshipped there as the god Nasr, and is mysteriously alluded to in Himyaritic inscriptions as 'the vulture of the east, and the vulture of the west,'[54] which also would seem to point to a solstitial use of the emblem. This important factor in the problem will be again referred to when we return in the last chapter to ancient Rhodesia.

CHAPTER XI

THE HIMYARITIC ROCK INSCRIPTIONS — EARLY SABAEAN AND MINAEAN CULTURE—ECHOES OF RHODESIAN CONTACTS.

MEANWHILE, we have to consider the actual relations of the early Himyaritic Hebrew and Phœnician seafarers in the Indian Ocean. A glance at the map will show that Moscha, the port of Ophir and chief emporium of the Sabaeans, and their Minaean forerunners, occupied a somewhat central position on the south coast of Arabia, nearly facing the island of Socotra, and well clear of the dangerous Cape Guardafui. It could not have been more conveniently situated for general trading purposes in the Indian Ocean, and as it was shown to be a Himyaritic port, it follows that this traffic was from the earliest times in the hands of the Arabs, and not of the Phœnicians, as is often assumed. At the opening of this enquiry I pointed out that the Himyarites were in fact the first in the field, and were only joined by their Phœnician kinsfolk somewhat later, that is, after Solomon had extended his sway over

Edom to the port of Ezion-geber, at the head of the Gulf of Akaba. In the national legends, confirmed again by Bent's explorations in the Bahrein Archipelago, their cradle-land was the west coast of the Persian Gulf, whence they reached their historic homes on the Mediterranean—Sidon, Tyre, Byblos, Berytus (Beyrut)—certainly not earlier than about 3000 B.C. Since that sufficiently remote epoch their migrations and colonia enterprise were exclusively confined to the Mediterranean and Atlantic waters (Leptis, Carthage, Carthagena, Cadiz, &c.). They were never heard of in the Indian Ocean, which, as we have seen, was at first an 'Erythraean Sea,' a 'Himyaritic lake,' so to say, till the second advent of the Phœnicians, the return to their primeval homes in the time of Solomon, about 1000 B.C. It was then that Hiram, King of Tyre, crossed into the Red Sea, apparently on the invitation of his friend and ally, Solomon, and equipped a fleet at Ezion-geber, either to join that of the Jews or to sail alone in quest of gold and the other treasures in the Indian Ocean. The texts are somewhat obscure, and in the authorised version even conflicting, for I Kings x. 22 tells us that 'the King had at sea the navy of Hiram,' as if he had chartered the Phœnician fleet for the purpose, whereas at II Chron. xxi. we read that 'the king's ships went to Tharshish with the servants of Huram,' as if the ships were his own, though perhaps manned by the 'shipmen' of Tyre. But the Vulgate comes again to the rescue, and in the earlier book [55] states distinctly that there were two fleets, the king's and Hiram's, and in the later Chronicles that the king's ships went to Tharshish with the servants of Hiram, as in the English text. It appears, therefore, that the king really had a fleet of his own, and we shall see farther on that at least some of the crew were Jews.

Meanwhile, two conclusions of great importance in this enquiry flow inevitably from these texts. The first is that the traffic in the costly merchandise was flourishing long before the advent of the Phœnicians in the southern ocean, because, as above pointed out, Solomon's father, David, had already accumulated a vast quantity of 'the gold of Ophir' (quite 1,000,000*l.* sterling) for the building of the Temple, while the auriferous Havilah itself is thrice mentioned in Genesis (chaps. ii. and x.), in the third instance in close connection with Ophir. The second conclusion is that this flourishing traffic was at first, that is before the time of Solomon and before the arrival of the Phœnicians, in the hands of the South Arabians, the Himyarites of Yemen and Hadhramout, consequently that the first arrivals in Havilah (Rhodesia) were not Phœnicians, but the Himyaritic Minaeans and Sabaeans.

The natural objection or even protest against this inference will, of course, be that in those remote pre-Phœnician times the Yemenites could scarcely have been civilised enough to possess seaworthy ships capable of navigating the storm-swept Indian waters. But the objection will be raised by those alone who are unacquainted with the true relations in South Arabia, relations which have only now been partially revealed by the decipherment of the already mentioned Himyaritic rock inscriptions. These inscriptions must now engage our earnest attention, the more so since it will be seen that the Himyaritic language, in which they are composed, has reached Madagascar, close to Rhodesia, while the script in which they are written has reached Rhodesia itself.

The documents in question, which since their recovery have proved 'a monument more lasting than bronze' for the Himyaritic nation, range over a very wide area, and although hundreds have already been copied and partly interpreted, probably many thousands remain still to be discovered. The script itself, often described as a modified form of Phœnician, reveals on the contrary a writing system more primitive than the oldest extant Phœnician, Moabite, or Hebrew characters, so old that Professor Sayce asks whether the Phœnician should not be derived from the Himyaritic rather than from the Egyptian hieroglyphs, or from Dr. A. J. Evans' recently discovered Cretan syllabary, as has also been suggested. In fact, Sayce shows that the forms of the Himyaritic letters afford a better explanation of the names of many in the Phœnician alphabet than do the hieroglyphs themselves. Thus *aleph* (the Greek *alpha*), meaning an ox, presents in Himyaritic the outline of an ox's head, but bears no resemblance to any of the Egyptian symbols used for *a*. Hence Sayce declares his present belief to be that the Phœnician alphabet, mother of all our European systems, 'came from Arabia.' 'Should this view be confirmed, South Arabian Semitic origins will have to be set back to a vastly remote period in order to allow time for the slow evolution of the Himyaritic script from the postulated pictorial prototypes to the already highly conventionalised forms of the oldest known rock inscriptions.' [56]

Such a hoar antiquity is partly revealed by these inscriptions themselves, which fall into two distinct groups—an early Minaean with fuller and more archaic Semitic forms, and a later Sabaean, though even this language is more primitive than that of the oldest Assyrian and Hebrew records. When we remember that the Phœnicians looked to the Persian Gulf as their cradle, that they must have been settled in the Bahrein Archipelago for long ages before their migration to the Mediterranean, and that Oannes, from whom the Mesopotamians received the germs of their culture, had also traditionally come up from the sea, further research may yet show that South Arabia was the source whence the Babylonians derived their first knowledge of the arts and letters. Even the cautious S. Lang suggests that 'the peculiar language and character of the race must have been first developed in the growing civilisation which preceded the ancient Minaean empire, probably as the later Stone Age was passing into that of Metal.' [57]

We have seen that the later Sabaean empire goes back with certainty to the time of Solomon, who was visited by Queen Balkis in Jerusalem; and Sabaea is shown by Assyrian documents to have still been a powerful state in the eighth century B.C., when it was conterminous northwards with the Ninevite empire under Tigleth-Pileser and Sargon III. But Saba (Sheba) was preceded by the far more ancient empire of *Mâ'in*, that is, of the Minaeans, whose very name had almost died out till rescued from oblivion by the lately discovered Yemenite stone records. These have already yielded a long list of no less than thirty-three Minaean kings, who were anterior to those of Saba, and ruled over the whole of Arabia as far as Syria and Egypt, as shown by the references to Gaza and to Teima (the Tema of Scripture) on the route between Sinai and Damascus. 'The kingdom of Sheba,' writes Sayce, 'arose after the decay of that of Mâ'in, or the Minaeans, and its rulers were already masters of Northern Arabia in the time of Tigleth-Pileser and Sargon (B.C. 733, 715). The Queen of Sheba had "heard of the fame of Solomon," for the northern limit of her dominions adjoined the southern limit of his' (*op. cit.* p. 65). Here is another proof, were it wanted, that Queen Balkis (Glaser writes *Bilkis*, from the inscriptions) did not cross the seas, but followed the very ancient caravan route from the capital to Jerusalem.

By the inscriptions as interpreted especially by Mordtmann, Müller,[58] and Glaser,[59] great light has been thrown upon the early records, religion, and commercial intercourse of the Himyaritic people. In Mordtmann much information will be found regarding the solar worship of the proto-Sabaeans, which appears to have originated with the Hamdán priestly order, of which most of the early Sabaean kinglets were members. The Riyám group worshipped Ta'lab, whose chief temple is at Riyám, north-east of Sana. I mention this because travellers penetrating into that inhospitable district should not miss the opportunity of exploring the remains of this edifice in connection with those of ancient Rhodesia. The flourishing traffic of South Arabia in spices is illustrated by the inscriptions on three incense-burners, where four distinct kinds of perfumes are specified, and here identified (p. 82). Here altogether fifty documents are examined, and in one of them (No. 9) reference is made to the Ahbashán people, who are not to be confused with the Abyssinians, but seem almost certainly to be the *Abaseni*, who are stated to have dwelt in an outer (foreign) auriferous land. Here we get a step nearer to the very name of the first (Himyaritic) settlers in auriferous Rhodesia. This name may yet be revealed on some of the inscribed stones of the Fura district south of the Zambesi, which Dr. Peters has seen and is now in search of.

Glaser has recovered over a thousand of these priceless records, which make it quite clear that in very remote times, coeval with, if not earlier than Babylonia and Egypt themselves, South Arabia, the Arabia of the children of Joktan, was 'a land of culture and literature, a seat of powerful kingdoms and wealthy commerce, which cannot fail to have exercised an influence upon the general history of the world.'[60] Everything points to Yemen as the Sheba of Scripture, which in the time of Solomon had extensive commercial relations with Tyre, and most probably with India and the East African seaboard, from Abyssinia to Sofala beyond the Zambesi. (Here I am giving Sayce's inferences from the rock inscriptions.) That the gold brought by the Tyrians and the Queen of Sheba came from Sofala (my Tarshish) and that it was the output of the neighbouring mines worked by the Semitic people of Ṣabaea, has been all but proved by Bent. Like the Egypt of Menes, the Sabaean empire was formed by the fusion of several petty theocratic states with that of the so-called 'Makârib,' or high priest of Saba, who gave his name to the land, as Assur did to that of the kindred Assyrian Semites of Nineveh.

From some of the above references it is evident that their sway was not confined to the mainland, and these proto-Himyarites, like their modern representatives, the Muhammadan 'Moors,' were great navigators, known in every port and haven in the southern waters. The Hon. A. Wilmot justly remarks that 'As the Phœnicians were the carriers of the northern world, so were the Arabs the marine purveyors of the southern ocean.'[61] There is good reason to believe that they taught, if not letters, certainly navigation, to the Babylonians, and Dr. Pinches (*loc. cit.*) tells me that 'in the cuneiform lists of parts of a ship occurs the term *kinnu*, which is the Arabic *khinn*, the *hold*.' He also gives me the forms *qarnu*, 'horn,' and *qarnati*, 'horns,' possibly meaning 'mast' and 'masts,' if not 'prow,' but in any case from the Arabic *qarn*, horn. In the same place he favours me with a veritable 'Homeric' catalogue of ships which has been published in the 'Cuneiform Inscriptions of Western Asia,' vol. ii., pl. 46, both in their Akkadian and Assyrian forms. Of great interest are several of these, such as *mairitum*=a Mairite ship, where *Mair* means 'the ship city,' or, as we should say, the naval station; *uritum*=a ship of Ur, where Ur is generally regarded as the 'Ur of the Chaldees'; *dilmunitum*=a ship of Dilmun, where Dilmun has been identified beyond reasonable doubt with the Bahrein Islands; lastly, *makkanitum*= a ship of Magan (Akad. *Maganna*), that is, the Sinaitic region. Here, therefore, we have both Babylonian and Arabian ships mentioned in the same early document, and, what is still more remarkable, the Cuneiform tablets record the fact that the merchants of Ur visited the distant Arabian lands here referred to, and that 'they encountered serious rivalry from the sailors of DILMUN and MAGAN, whose maritime tribes were then, as now, accustomed to scour the seas.'[62] Some Assyriologists regard these ancient craft as mere coasters confined to the Persian Gulf. But others (Boscawen, Brindley, Oppert) show that they were stout enough to navigate all the Arabian waters, and bring back great blocks of diorite from Sinai. Dr. Pinches also quotes me a letter seemingly addressed to Ashur-banî-âpli, in which the writer informs the king that 'Ashshur-mukin has aided me. I caused to ride in the ships the stone of the bull (and) the Colossus. The ships were not (too) deep—they did not ground, &c.' (*loc. cit.*).

Navigation both in Arabia and Babylonia is thus set back to Akkadian times, and we now know that both the Akkadian and Arabian (Minaean) records date from several thousand years before the new era. But no materials have hitherto been available, or at least utilised, to determine the southward range of the Himyarite seafarers in the Indian Ocean. These I can now supply, and hope by their means to show that the traders of Ophir penetrated southwards to Madagascar, and thence across the narrow Mozambique Channel to Rhodesia.

CHAPTER XII

THE HIMYARITES IN MADAGASCAR

HERE we enter a field which in this connection may truly be called virgin soil. The facts which I shall now produce are not indeed quite new, and some have been known for over a quarter of a century. But they have never been studied in their bearings on the subject now under consideration, although one or two vague references have been made to them since the exploration of the Rhodesian monuments began to attract attention. These facts therefore call for more ample treatment than was found necessary in dealing with some of the older phases of the Ophir question.

It is generally assumed that Madagascar was unknown to the ancients, and in a sense that is true. Commenting on the wooded island of Menuthias briefly described by the Periplus (§ 18), Tozer remarks (p. 275) that, 'according to the distances that are given, this must correspond either to Zanzibar or to the more northerly island of Pemba.' So also Glaser (p. 206): 'The distance suggests Zanzibar,' but adds: 'while the great rivers point to Madagascar.' Now the 'distances' where the early writers speak, as they do here, at second hand, literally count for nothing. They had no notion of the actual size of the globe, and even Columbus, when he first struck land, thought he had reached the East Indies, an error which is perpetuated by the expression 'West Indies,' and by the term 'Indians' applied by him to the aborigines of the New World. Glaser continues: 'So much is established that at that time Himyar possessed [politically] nearly the whole of East Africa, and in those gunless and powderless days such an extension of sway implied long centuries of efforts.' The significance of this statement, derived from the rock inscriptions, will appear later on.

At this part of the narrative the Periplus makes the unexpected remark that the ocean, although unexplored, was known to trend round to the west beyond Menuthias, and eventually mingle with the waters on the other side of Africa. This does not fit in with Zanzibar at all, but does with Madagascar, and shows conclusively that the terminal point of the continent was at that time known at least by report. The knowledge might have been handed down traditionally from the days of Necho's expedition. But six or seven centuries are a long period for such a tradition to survive, and it seems much more probable that the information was obtained directly from the Himyarites, who had for ages ruled over the eastern seaboard.

Apart from this consideration, the detailed account of Menuthias excludes Zanzibar and Pemba altogether and compels us to accept Mr. J. T. Last's identification with the great island away to the south : 'Shortly after the commencement of the Christian era two navigators, Theophilus and Diogenes, passed Cape Guardafui, and reached the port of Rhapta, by some thought to be the same as the present Zanzibar, but more probably a point near Mozambique in about 15° S. Soon afterwards the pilot Dioscorus doubled Cape Guardafui, and, proceeding still further south, arrived at Cape Prasum in about 22° S., near the mouths of the Zambesi. . . From the description given by Arrian, there is much reason to think that when he speaks of the island of Menuthias, he is referring to Madagascar. His remarks would scarcely apply to any other island off the East African coast ; his description of the [great] rivers, crocodiles, land-tortoises, canoes, sea-turtles, and wickerwork weirs for catching fish, apply exactly to Madagascar of the present day, but to none of the other islands.' [63]

But it may be argued that if Madagascar was so well and so long known to the Himyarites, the Greeks and the Romans must have also been well acquainted with it. To this there is a complete answer, which is very well put by Wilmot (*op. cit.* p. 90) : 'If there be any feature more striking or remarkable than another in Arabian history, it is the extreme secretiveness of the nation. Ten centuries after the vessels of the Tyrians and Syrians had ploughed the southern seas, we are told of the wonderful discovery of monsoons in the Indian Ocean by Hippalus. Of course they were known to the Arabians during all that period, and naturally they used the knowledge, and benefited by it. Then what extreme reticence must they have preserved as to the situations and character of the places with which they traded. We have seen already that, according to Strabo, the master of a Phœnician bark preferred to lose his vessel on the rocks of the Cassiterides rather than permit a Roman ship which followed to gain a knowledge of a safe channel to a valuable port. The secret of gold was kept quite as carefully as any other, and for even more powerful reasons.' Thus it happened that, apart from a curious and erroneous, or at least misunderstood, allusion by Marco Polo,[64] nothing was known of this strange fragment of a vanished Indo-African continent till the advent of the Portuguese in the eastern seas. There are clear indications that the 'Land-under-the-sky,' as the natives call it, has been occupied by man from extremely remote times, possibly in the Old Stone, but certainly in the New Stone, Age. Whether the first arrivals were Bantus and other Africans from the neighbouring continent, or Oceanic Mongols from distant Malaysia, cannot now be determined, but that both races have been present throughout historic times, and are everywhere blended in diverse proportions, is no longer seriously questioned. The remarkable fact which here more nearly concerns us is that, despite the various racial types, there is but one linguistic type, the Malagasy language being everywhere exclusively spoken with comparatively slight dialectic divergencies. It has no kind of affinity with Bantu, Hottentot, or any other African tongue, but is a recognised member of the widespread Oceanic family commonly known as Malayo-Polynesian. It might be supposed to be most closely connected with the standard Malay of the Eastern Archipelago; but that is not so, and it presents far more archaic forms than does that relatively modern member of the group. Malagasy had been established in the island long before the historic Malays had begun to spread abroad from their original seats in Sumatra, long before the Malay lands had been reached over 2,000 years ago by the Hindu missionaries from India. This is shown by the absence of Sanskrit words introduced by those missionaries in considerable numbers into Malaysia, but not one of which has found its way to Madagascar. The Oceanic migrations to this island were therefore prior to the spread of Hindu influences in Sumatra, Borneo, Java, and other parts of the Great Archipelago.

When the statement that no African idiom was current in Madagascar, where an Oceanic tongue was everywhere exclusively spoken, first appeared in 'Rhodesia' (September 21, 1901), it was received with incredulity by a learned critic, who inferred from the proximity of the island to the mainland, from an admitted widely diffused strain of black blood amongst the natives, and so on, that their speech *must* also be largely or mainly of African orgin. Naturalists are aware how faulty this sort of reasoning would be if applied to the very peculiar indigenous flora and fauna. It is even more faulty when applied to the Malagasy language, for the absolutely Oceanic origin of which overwhelming evidence will now be submitted.

Linguistic kinship is established by the test of vocabulary and structure, and in these respects Malagasy shows no kind of likeness with the African, while its agreement with the Oceanic family is complete. There are scarcely any African terms to be found anywhere, and even these are merely loan-words mostly introduced by the Arabo-Swahili traders.[65] Such are *kanga*, Mal. *akanga* = guinea-fowl; *ng'ombe*, Mal. *omby* = cattle; *ntango*, Mal. *voa-ntango* = gourd; *mbwa*, Mal. *amboa* = dog; *mananazi*, Mal. *mananasa* = pine-apple; (the European *ananas*, originally an American (Guiana) word). Of African (Bantu or other) structure, such as the alliterating pronominal prefixes, there is absolutely no trace, so I can give no Malagasy analogies.

The lexical affinities with Malayo-Polynesian are countless, and here it is to be noticed that many of the Malagasy terms are lost in Malay, but recur elsewhere in the Indo-Pacific archipelagoes, the explanation being that these terms belonged to the common original Oceanic stock, and were carried with the primeval migrations east to the Melanesian and Polynesian islands, and west to Madagascar. Such are:—

English	Malay	Malagasy	Oceanic
ship	prau	lakana	lakatoi (Motu, N. Guinea)
body	badan (Arab)	tena	tinana (Maori, New Zealand)
hair	rambut	volo	vulu (Fiji)
head	kapala (Sanskrit)	loha	ulu (Samoan) lova (Vaturana, Solomon Islands)
mother	ma	reny	reme (Torres Islands)
mouth	mulut	vava	vava (Espiritu Santo, New Hebrides)
rain	hujan	ranonorana	rani (S. Cristoval, Solomon Islands)
soft	lumbut	malemy	malumu (Savo, Solomon Islands)
tongue	lidah	lela	alele (Rotuma, Mid-Pacific)
tooth	gigi	nify	nifo (Samoan)
nose	hidong	orona	urong (Dayak)
tree	kayu	hazo	hasie (S. Cristoval)

English	Malay	Malagasy	Oceanic
woman	parampuan	vehivavy	vavine (Oba, New Hebrides) fafine (Samoan)
ear	chuping (lobe)	tading	taringa (Easter Island)

Here are a few terms, which besides their immense range—from Madagascar to Tahiti, and even Easter Island, within measurable distance of South America—are otherwise specially interesting :—

In the Hova dialect *sun* is *maso-mahamay* = 'burning-eye'; but elsewhere in Madagascar, it is *maso-andro* = 'day's eye,' as in Malay (*mata-ari*), Bajau of Celebes (*mata-lon*) ; Salayer (*mato-allo*), &c.

The Malagasy *mulutra*, lips, is the Malay *mulut*, mouth, a curious instance of a part for the whole.

The Javanese *ati* means both 'heart' and 'liver'; but the Malag. *aty* 'liver' only ; yet a royal prince is *ati-n-andriana*, 'liver-lord.' This is explained by the Bisayan (Philippines) *atay* = 'liver' and 'courage.' Thus : *dako ang atay-nia* = 'he has a great liver.' *Cf.* English *white-livered*. This word reaches all the way to Easter Island, where *até* = 'liver.'

The Malag. *vity*, 'leg,' explains the Kayan (Borneo) *biti* = 'to stand up.'

The Malag. *ranu-maso*, 'tears' = 'eye-water,' like the Javanese *ranu-moto*, Malay *ayer-mata*, and Samoan *loi-mata*.

The Malag. *voa-vity*, calf (of the leg) = 'fruit of the leg,' as in Malay *buah-betis* ; Javan. *woh-wentis*, &c.

The Malag. *vovu-masu*, 'eye-lid,' is lit. 'eye-roof,' as in Tagalog; but the Malay is *kelupak-mata*, 'eye-lid,' as in English.

The Malag. *vurona*, bird, is explained by the Kavi (Old Javanese) *búr* = 'to fly'; while *en revanche* the Javan. *ikan-chuchut*, chuckut-fish (shark) is explained by the Malag. *suta-suta* = 'glutton.' Again, Malag. *valála*, 'locust,' is explained by Macassar (Celebes) *balála*, which also means 'glutton.' Thus there is a continual give and take between these Oceanic tongues, which everywhere supplement each other, often in the most unexpected way. The Malag. *lelafu*, 'flame,' is a contraction of *lela-afu* = 'tongue of fire,' like the Malay *lidah-api*. But the Javanese say *urub*, from Malag. *uru*= 'to burn.' Lastly in Malaysia the fishing-net is the Sanskrit *jala*; but the old word was *saratu* or *haratu*, and this alone occurs in Malag., in which there are no Sanskrit elements. This again shows that Madagascar was peopled from the Eastern Archipelago in pre-Hindu times, as it was brought into relation with the Himyarites in pre-Muhammadan times.

Coming now to the numerals, we find that the primitive Oceanic names of the units up to ten included survive in a great number of Malayo-Polynesian tongues, scattered over the two oceans. But four of them (3, 7, 8, 9) are lost in Malay, and not one in Malagasy. Thus:

	One	Two	Three	Four	Five
OCEANIC . .	sa	rua	tol	pat	lima
MALAGASY . .	isa	rua	telu	apat	limy
MALAY . . .	sa	dawa	tiga	ampat	lima

	Six	Seven	Eight	Nine	Ten
OCEANIC . .	an	pitu	walu	sio	pulu
MALAGASY . .	anam	fitu	valu	sivy	fulu
MALAY . . .	enina	tujuh	dalapan	salapan	puluh

This is startling, and seems almost incredible ; but it is to be remembered that the Oceanic, like the Semitic (see above), is characterised by an astounding vitality and persistence which is quite foreign to the Aryan and all other linguistic families.[66] In the Semitic it is the consonants, as we have seen, in the Oceanic the vowels, that show most tenacity, the former being like the sturdy oak that resists the storm by its inherent strength, the latter like the reed that bends to the gale.

Let us now take one or two of the structural elements, *tafa*, for instance, which is a prefix expressing spontaneous action, used something like our English *self*, in such combinations as *self-help*, *self-evident*, &c. It appears to prevail in various modified forms—*tav*, *ava*, *tapa*, *tara*, &c.—over the whole Oceanic area, and 'the resemblance between the Malagasy *tafa* and the Banks Islands *tava* is so complete in form and signification, and this in a fine point of meaning, that, considering the space of ocean that separates the languages [the Banks group lies north-east of the New Hebrides], it is a matter of astonishment that it should exist. It is impossible that it should be accidental ; it could not be introduced [into Madagascar] by Malays or Polynesians who have it not ; it must have survived no one can tell what vicissitudes and changes, in a course of years no one can number, and presents itself, like a rare species of plant or flower, in isolated and widely separated localities, a living and certain proof of common origin and kindred' (Last, p. 60).

Lastly a peculiar feature of the Oceanic system are the infixes, that is, modifying particles, not agglutinated as in the Mongolo-Turki family, or prefixed or suffixed as in Aryan, Bantu, and many American groups, but embedded in the very body of the word. They are wonderfully uniform, being mostly *in*, *on*, *om*, also *mn*, *l* and *r*, and range sporadically, as I have elsewhere shown,[67] from Indo-China, and the Philippines, through the Malay lands to Madagascar. Thus, in Khmêr (Cambojan) *chereap* = to know; *chumreap* = to make known, to teach; *kur* = to draw; *komnur* = a drawing, a design; *sauk* = to corrupt; *samnauk* = a bribe; in Tagalog (North Philippines) *tapay* = to knead, *tinapay* = bread; in Malay *pukul* = to strike, *pâmukul* = a hammer; *sipit* = to grasp; *sinipit* = an anchor; in Malagasy (here very numerous) *tady* = twisted, a rope; *tomady* = strong; *safotra* = overflown; *somafotra* = brimful; *safy* = spying, *somafy* = sight of a distant object; *sany* = likeness, *somany* = like; *sisika* = forced in, *somisika* = shattered.

This perhaps will satisfy my sceptical critic, who, if still doubtful, might here be asked to produce a thousandth part of the evidence in support of his theory that I have now advanced in support of mine. But he cannot do so; nobody can, for it is non-existent and *ex nihilo nihil fit*.

It was above pointed out that in Malagasy there are no Sanskrit elements, the presence of which would argue a comparatively recent (post-Hindu) arrival of the Oceanic peoples from Malaysia. But there are numerous Semitic elements, and those of a nature to argue very ancient and long-continued contacts with Semitic peoples from Arabia. These Semitic elements are usually spoken of as Arabic, and are attributed to the

Moslem Arabs, or to the Moslem Swahili people, who have settlements on the north-west coast dating back a few hundred years. Some are undoubtedly due to this source, but others, and those the more numerous and important, date back, not a few hundred but a few thousand years, and are due, not to the Moslem Arabs of later times, but to the Himyaritic Arabs of the Sabaean and Minaean epochs. Such terms, as might be expected, are of a more or less cultural nature, and are concerned with soothsaying, divination and certain early Semitic rites as well as with the divisions of time —the seven days of the week, the twelve months of the year—which the Himyarites had in common with the Babylonians, but of which the early Oceanic peoples had no knowledge. Here we reach a turning point in the discussion, and the proofs which I now submit that the Malagasy names of these divisions of time are not derived from the modern Arabic of the Koran, but from the ancient Arabic of the rock inscriptions, will also prove that these southern latitudes at the very gates of Rhodesia were first reached, neither by the Phœnicians nor by the Muhammadan Arabs, but by their Sabaean and Minaean precursors. In Malagasy, Himyaritic, the Arabic of the Koran and Malay, the weekdays are as under :—

English	Malagasy	Himyaritic	Neo-Arabic	Malay
Sunday	Alàhady	Al-àhadu	El-àhad	Ahad
Monday	Alatsinainy	Al-itsnàni	El-etnèn	Sinnen
Tuesday	Tàlata	Tsàlatsatu	El-t'late	Salata
Wednesday	Alarobìa	Al-arbàata	El-àrba'e	Robu
Thursday	Alakamìsy	Al-khamìsu	El-khamìs	Kumis
Friday	Zoma	Juma'tu	El-Juma'	Jumat
Saturday	Asàbotsy	As-sàbtu	Es-sàbt	Saptu

Thus nearly all the Malagasy words are seen to represent the old Himyaritic, while the Malay, borrowed in recent (Muhammadan) times, are derived from the Neo-Arabic of the Koran. The inference is inevitable, and it must be clear to everybody that had the Malagasy people taken their word for Sunday, to quote one instance, from the Neo-Arabic, it must have presented some shortened form like the Malay *ahad*, and could not, short of a miracle, have harked back to the proto-Arabic (Himyaritic) *alàhadu*. Linguistic are not like animal and vegetable species, which evolutionists tell us are liable to reversion. The people therefore who brought these terms into the island were not recent Arabs, but ancient Arabs of Himyaritic speech, and we have above shown that Himyaritic is the most archaic of all the Semitic tongues. I lay stress on the point, because it is so difficult to induce students wedded to old ideas to accept new views, especially when these are of a subversive nature. They wrestle with the facts, and when overthrown prefer still *stare per vias antiquas* rather than yield to the self-evident, and I think this is a strong trait in our conservative English temperament.

But if not yet convinced, they may perhaps be converted by the names of the months, which 'are not those of the Arabic solar or lunar months, but those of the twelve signs of the Zodiac, of Sabaean or Babylonian origin, and these are used in many parts of Madagascar, not only as the names of the month, but also as names for the several days in each month, that is, using four of those names for three successive days each, and the rest for two, always in the same order in which the names are taken to follow one another in the Zodiac.'[68] The process was analogous to that of the Mexican (Aztec) *Tonalamatls* (series of lucky and unlucky days), and the explanation may be the same—the necessity of having names and signs for each day as well as for each month with calendric systems used for casting horoscopes and such like purposes. As in Babylonia and Sabaea, these signs and names were from the first, and still are, used in Madagascar in connection with the *vintana*, that is, the good and bad days, and the *sikidy* or *sikiry*, that is, the 'medicine,' magic, or counter-charms against the evils threatened by a bad vintana. This very word *sikiry* is the Chaldaean *zekuru*, whence also the rare Greek word *zakoros*, a priest or wizard. All the Malagasy names of the month (signs of the Zodiac) are similarly of old Semitic origin, and, like the names of the week days, must have been introduced by the Himyarites long before the Christian era. They are :—

1. Alahamady ...	Sem. Al-hamalu	the Ram
2. Adaoro ...	Sem. At-tauru, Chal. Thor, Syriac thauro	the Bull
3. Adizaoza ...	Sem. Al-jehauza'u	the Twins
4. Asorotany ...	Sem. As-saratanu	the Crab
5. Alahasaty ...	Sem. Al-asadu	the Lion
6. Asombola ...	Sem. As-sunbulu ... ... properly *Spica* in *Virgo*.	the Virgin
7. Adimizana ...	Sem. Al-mizanu	the Scales
8. Alakarabo ...	Sem. Al-aqrabu	the Scorpion
9. Alakaosy ...	Sem. Al-qausu	the Archer
10. Alijady ...	Sem. Al-jehadia	the Goat
11. Adalo ...	Sem. Ad-dalwu	the Water-bearer
12. Alohotsy ...	Sem. Al-hutu	the Fishes

Thus, allowing for the different Semitic and Oceanic

phonetic systems, and for the normal interchange of *l* and *d*, *t* and *ts*, &c., the correspondences are complete. In any case it is obvious that these Babylonian names of the Zodiacal constellations could not have been introduced into Madagascar as the names of the months by the later post-Koranic Arabs, who would have brought their own proper calendric nomenclature, this being required for determining and regulating the feasts and the fasts of the Muhammadan religion. Hence Mr. Last, referring to the non-introduction of this religion by the Arabs, though they instructed the Malagasy people in their pagan system of astrology and divination (*ventana* and *sikidy*), rightly observes that 'this seems to imply that the first immigration of Arabs [Himyarites] took place . . . before Arabia and the surrounding countries had accepted that religion' (*loc. cit.* p. 62), that is, as here contended, in pre-Muhammadan times.

But are these Himyaritic terms really the current names of the months throughout the island, or are they, as has been lately suggested, merely the esoteric names jealously preserved for divining purposes by the initiated, perhaps in the south-eastern district? This district is known to have been visited and even partly settled by some Arab immigrants at an early date. 'The Arabs who came there had brought their all with them, and they had come to stop. They probably came in considerable numbers, for nearly all the south-east tribes claim their descent from them; and that they gained an ascendency over the natives, either by force or superior skill, there can be no doubt, for in the course of time their permanent residence in the island *affected the whole social system of the Malagasy*. In this part of Madagascar we see much that would lead people to think that a relationship existed between the Arabic and Malagasy languages. This might at first sight seem to be indicated by the considerable number of words which have been introduced by the Arabs who came and settled in this part of Madagascar; also by the fact that *many of the natives, from ages past down to the present time*, have been able to write and read a kind of Arabic, and also to decipher *a number of old Arabic manuscripts* they have in their possession, and which they guard with the most jealous care' (Last, p. 54).

Of course the relations here spoken of as possibly existing between Malagasy and Arabic are purely lexical, in no way affecting the structure of the former or weakening its claim to membership with the Oceanic family, of which it remains a typical branch. The case is somewhat analogous to that of English and Latin, with this difference, that for one Semitic word in Malagasy, there are a thousand Latin (and French) words in English, without depriving it of its right to be regarded as a member of the Teutonic group.

Now, this 'kind of Arabic,' is it Himyaritic or Koranic? And, above all, the manuscripts, which are so often heard of, but have never yet been seen by any European, if they exist at all, in what script and language are they written? If in Himyaritic, about which I am more than sceptical, then their recovery would be worth a trip to Madagascar several times over. A specimen of Sabaean penmanship would indeed be a rare find with which to open the twentieth century, a find eclipsing most of the palaeographic discoveries of the last two centuries.

But for our present purpose we have still to ask, do these documents contain the astrological lore above described, and is this lore confined to these documents so jealously guarded by their owners? To this I can unhesitatingly reply that the lore is independent of the documents and their interpreters, is diffused over the whole Malagasy domain, has become a national inheritance since its introduction by the Himyaritic star-gazers thousands of years ago. The evidence for this is supplied especially by the English and Norwegian missionaries—Cousins, Sibree, Dahle, Richardson, Pearse, and others—who have visited and associated with all the chief tribes or nations, and thus collected at first hand a great mass of materials, showing that the calendric systems, religious practices, superstitions, social usages and traditions have everywhere been influenced by Semitic peoples, as seen not only by analogies and parallelisms, but also more directly by the numerous Semitic verbal elements permeating them. The Malagasy social life is saturated by Semiticism, just as the neighbouring Rhodesian uplands are honeycombed by Semitic gold-workings and strewn with the remains of Semitic monuments.

The above-quoted table of Himyaritic month-names does not occur uniformly everywhere, being prevalent chiefly amongst the dominant Hova people of Imerina—the north central tableland.[69] But Semitic formative and other elements crop up everywhere, as, for instance, in the Betsileo names of the eighth, ninth, tenth and eleventh months—Asàra, Asàramanàra, Asàramànitsa, and Asotrizonjona—where the first part of the name (*asar*) is the Arabic *al-sahr* (pronounced as-sahr)=' the month.' So also Sàramàntsy and Sàramànitra, the eighth and ninth months of the Tanala nation.

It was above seen that the month-names serve also generally as day-names; but that is not so in some of the southern districts, where there are different names for the days. All these names are now found to be also Semitic, that is the old Arabic names of certain conspicuous stars in the several constellations from which are taken the names they have in common. Thus in the first month *Alahamady* (see above) the first three days are all called Alahamady, but taken singly the first two are called *asoratin* and *alabutin*, the former being the Arabic *ash-sheratain* = β and γ Arietis, the latter also the Arabic *albotein* = δ Arietis, and so on for all the twenty-eight days of the lunar month. The adaptations to the Malagasy phonetic system are often peculiar, making the identifications difficult. Thus the familiar *aldebaran* α Tauri is disguised as *adobora*; *az-zubra* (o Leonis) as *hazubura*; *az-zaben* (α Libræ) as azubana, &c. It is certainly a revelation to find the Sabaeo-Babylonian astronomic nomenclature still surviving amongst the unlettered and semi-barbarous Oceanic populations of Madagascar. When we read the fierce attacks often made on the Evangelists amongst the lower races we should at least remember these discoveries, which have been made by them, and throw so much fresh light on the obscure relations of the early peoples in the eastern world.

Dahle thinks the word *vintana* itself (see above) may be Semitic. 'Vintana is the destiny, especially as depending on time. To be born under a bad vintana is to be born on an *unlucky* day. Now the Arabic *evan*, plur. *evinat*, means *time*, especially time viewed as fit or unfit for what is to take place in it, or seasonable. This comes very near to the Malagasy *vintana*. Although it was a miserable fate to be born on an unlucky day, it could to a certain extent be remedied or counter-charmed by the Sikidy. Sikidy has generally been translated "divination"; but its meaning is a wider one, including also the whole practice of pointing out powerful counter-charms against almost every kind of evil, but especially those originating from a bad vintana or from witchcraft, the two supposed sources of nearly all misfortunes in Madagascar. . . . Out of the three different tables of names that are necessary in the practice of it [Sikidy] the first one, containing sixteen names, appears to be *entirely Arabic*, whilst in the two others Arabic names are mixed up with Malagasy ones' (*loc. cit.* p. 80). In a word a large measure of the ancient Sabaeo-Babylonian astrological conceptions, especially those associated with divination, are found to be still surviving amongst all sections of the Malagasy people.

Although they came of a famous seafaring race, they appear after their arrival to have forgotten the art of navigation. By following the marine current setting from north-east to south-west, it was not difficult to pass from island to island, and thus reach the east coast of Madagascar. But the return voyage was scarcely practicable, and after the first settlements all communication with the Malaysian homeland seems to have been discontinued. A few Malay corsairs or traders may have occasionally arrived in later times; but not in sufficient numbers to exercise much influence over the natives, as shown by the non-Malay character of their Oceanic speech. The Indian Ocean had meantime become a 'Himyaritic lake,' and these South Arabian populations did not need to be taught how to navigate these southern waters by the Malayo-Polynesian seafarers. We have seen that they were rather the teachers even of the Babylonians themselves, as they most probably were also of the Malagasy. This may account for the absence of the Malay word for boat (*prau*), which is replaced by the Arab *sambuk* (*sambo* through the Swahili *chambo*), and it may be remembered that it was the Arab sambuk and not the Malagasy sambo or the Malay prau that the Portuguese pioneers found plying along the east African seaboard or scouring the high seas. Thus Barbosa tells us, speaking of Sofala, that in his time (about 1512) 'it was a town of the Moors [Arabs], who established themselves there a long time ago on account of the great trade in gold, which they carry on with the gentiles of the mainland. And the mode of their trade is that they come by sea in small barks which they call *Zanbucs*, from the kingdoms of Quiloa and Mombaza and Melindi,' &c. These places follow along the east coast northwards in the direction of Somaliland, where the same traveller found the squadrons of his Portuguese fellow-countrymen lying in wait 'about Cape Guardafun [Guardafui] for the [Arab] sambuks plying between the Red Sea and India, to take them with all their riches.'

There is nowhere mention of any Malay craft frequenting the Indian Ocean west of the Eastern Archipelago, and all the historic evidence goes to show that till the arrival of the Portuguese these waters had been exclusively navigated by the Babylonians and Semites [Himyarites, and for a short period their Phœnician and Jewish allies] since prehistoric times. I cannot therefore follow the reasoning of Mr. Syme, who in 'Rhodesia' (*loc. cit.*) argues that 'this pre eminently maritime race [the Malayo-Polynesians] had extended itself from isle to isle more than half round the world—from Madagascar through the Malay Archipelago, across the Pacific nearly to South America. It is probably due to them that navigation flourished round the coasts and on the rivers of India and Ceylon, and that *it was introduced by them to the peoples on the Persian Gulf and South Arabia.* The Persian Gulf is commonly regarded as the cradle of navigation, but it is not likely to have been so. One of the essentials of such an art was lacking, viz. suitable timber for ship-building. This of course was plentiful in India, and in the Malay Archipelago, as well as in Madagascar. Therefore in all probability the proto-Malays long held the monopoly in ocean traffic until the Semitic peoples turned their energies in this direction.'

This is a very good specimen of the *à priori* argument, which, however, must yield to facts. One material fact is that, as above seen, the proto-Arabs and Babylonians had forestalled the proto-Malays in securing a monopoly of the sea-borne traffic in the Indian Ocean long before the Malays proper were ever heard of. Another important consideration is the character of the climate, which eight or ten thousand years ago was certainly far more humid than at present, or throughout the late historic period. At that time there was an abundance of suitable timber on the then well-wooded Gara (Dhofar) and North Somali coast ranges, in some parts of Hadhramout, almost everywhere in Arabia Felix, and probably also round the shores of the Persian Gulf, just where it was wanted. A clump of those 'great oaks' alone, in the thick boughs of one of which Absalom 'was taken up between the heaven and the earth' (2 Sam. xviii. 9) would have sufficed to build quite a little squadron of the light craft of those days. Thus is put aside the timber difficulty, and no other presents itself. The sway of the South Arabian Himyarites over the high seas, their range southwards to Madagascar, and their long contact with the Oceanic inhabitants of that island are now, I hope, established beyond reasonable doubt.

CHAPTER XIII

THE JEWS AND PHŒNICIANS IN MADAGASCAR

I HAVE above suggested that the fleets of Hiram and Solomon must have been partly manned by Jews. It seems impossible otherwise to account for the numerous early Jewish as well as other Semitic rites and usages still everywhere prevalent in Madagascar. Of course we all know that after the destruction of Jerusalem by Titus (70 A.D.) there was a general dispersion of the Jews, many of whom took refuge in various parts of Arabia and especially in Yemen, where they even founded some petty states whose rulers—Asad Abu-Karib, Dhu Nowas, and others—are still remembered.[70] But long before this time intimate relations had been established between the still united people of Israel and their southern Himyaritic cousins and neighbours. Such was the fame of the Solomonic and Phœnician expeditions to the southern seas in quest of treasure, that earlier and far from unsuccessful essays in the same direction appear to have been forgotten. Else it would have been remembered that Solomon's father, David, had already led the way, and thus obtained that abundance of the precious metals to which reference was made in Chapter IX.

Dr. Peters calls attention (*op. cit.* p. 54) to a fragment of Eupolemos preserved in Eusebius (*Praep. ev.* ix. 30) 'which relates that King David sent miners to an island called Urphe, according to Gesenius more correctly called Οὐφρή or Οὐφήρ, on which were many gold mines, who brought gold thence to Judaea. It will be conceded that this fact of mining operations must speak against the position of Ophir in India [assuming of course that David's Urphe was Ophir], for the Aryan tribes, which in those times just then began their career of conquest, were not at all likely to have suffered any military occupation for the protection of mining undertakings. If, therefore, we maintain that the Ophir gold was got by mining, we must look for Ophir in regions inhabited by less valorous [or less civilised] races. These, however, were not to be encountered on the eastern, but on the western shores of the Indian Ocean.' And again (p. 113): 'It follows with certainty [that is, from the mention made by David in 1 Chron. xxix. 4 of his 3000 talents of 'gold of Ophir'] that David already had relations with Ophir.'

I have introduced these passages, not to re-open the question of Ophir,[71] which I trust has been sufficiently discussed in previous chapters, but in order to recall the undoubted fact that David really did anticipate Solomon in despatching expeditions for foreign treasure. Now here two considerations present themselves. The first is that such expeditions must have entered the Himyaritic political domain, because, as Dr. Peters here rightly points out, the auriferous land must in any case have been sought, not on the east, but on the west, side of the Indian Ocean; that is, on the East African seaboard, nearly the whole of which Glaser tells us (see above) was politically held by the Himyarites for a very long period of time, and certainly when the South Arabian empire was flourishing under the Queen of Sheba.

The second point to be considered is that for the greater part of his reign David was engaged in continual warfare with the Philistines and other Canaanites, whence Solomon's message to Hiram that 'David my father could not build an house unto the name of the Lord his God, for the wars which were about him on every side, until the Lord put them under the soles of his feet' (1 Kings v. 3). No doubt it is stated at v. 1 of the same chapter that 'Hiram was ever a lover of David.' But it is impossible to suppose that any formal alliance had been contracted between these two potentates, else mention would be made of the circumstance in 2 Samuel, where the fullest account is given of the 'Man of Blood,' as he is called in the Talmud. Hiram, indeed, sent him gifts of 'cedar-trees' to build him 'an house' (2 Sam. v. 11), that is, no doubt, the Temple, as shown by the corresponding passage in 1 Chron. xxii. 4: 'And cedar trees in abundance: for the Zidonians and they of Tyre brought much cedar wood to David.' But there was no concerted action of any kind, least of all a gathering of the fleets in the Red Sea, to which Hiram had not yet gained access. But it is important to note that after slaying twenty odd thousand Syrians and reducing Damascus, David spread his rule southwards, and 'put garrisons in Edom, throughout all Edom put he garrisons, and all they of Edom became David's servants' (viii. 14). This brought the Israelites down to the Red Sea, because Edom (Idumaea) was that debatable land between Canaan and Arabia, which extended from Palestine to the head of the Aelanites Sinus (Gulf of Akaba). This gulf was so called from the ancient city of Aelana (Ailath), just north of Ezion-geber, whither Solomon afterwards came. Then it was that they began to talk of 'the coasts of Israel,' although the expression may at first have been taken in a somewhat elastic sense;[72] for Israel had now at last obtained access to the seaboard with the road open all the way to the auriferous lands of the southern waters. And David's people had clearly gone forth on these high seas, probably in company with their Sabaean kindred, since the Phœnicians had not yet arrived from the Mediterranean. Hence the vast store of treasure, the fame of which seems to have waxed stronger with the years down to later times, when the Chronicler, writing after the Captivity, makes David say to Solomon: 'Now, behold in my trouble I have prepared for the house of the Lord an hundred thousand talents of gold, and a thousand thousand talents of silver, &c.' (1 Chr. xxii. 14).

The Israelites were thus, before even the advent of the Phœnicians, brought down to Menuthias, the great island of Madagascar, where, like the Himyarites, they have left undying memories of their presence, as we shall now see. The later Jews referred to at the opening of this chapter had long discontinued such primitive and anthropomorphic practices as those associated, for instance, with the scape-goat of Leviticus xvi. 10: 'But the goat, on which the lot fell to be the scape-goat, shall be presented alive before the Lord, to make an atonement with him, and to let him go for a scape-goat into the wilderness.' And verse 22: 'And the goat shall bear upon him all their iniquities unto a land not inhabited; and he shall let go the goat in the wilderness.' Now it is precisely these ancient ritualistic observances that we find surviving in Madagascar. The goat, not being indigenous, is replaced by the ox, and the late James Cameron, to whom I am indebted for several of the following incidents, mentions the case of a man doomed by his destiny to a violent death who was told by the priests that the calamity might be averted by doing as follows. 'He was to ride on the back of a bullock, and to take in his hands a small calabash filled with blood. He was then to pour out the blood upon the head of the bullock, then to dismount and send the bullock far away into a desert, where it should never more be seen or heard of.' Sad to say, after scrupulously carrying out all the instruction, the doomed person met his bad vintana all the same.

The same notion of vicarious sacrifice, common enough amongst the Hovas of Imerina, is illustrated by the case of King Andriamàsinavàlona, who when at death's door is told by the priests that he may be saved if a man can be found willing to die for him. In response to a proclamation inviting offerings most of the people ran away, but one, a certain Trimofòloàlina, presents himself with the gullet of an ox filled with blood concealed round his neck. This the priest slits, blood is shed, the gods are tricked and appeased

Amongst other striking analogies noted by Cameron are the following.

Many of the early Israelites combined the worship of the Deity with that of the sun, moon, and stars. The natives of Imerina do precisely the same thing. In seeking blessings for themselves and others, their prayers are addressed to God the Creator, the great orbs and the stars indiscriminately.

Amongst the ancient Hebrews provision was made for the detection and punishment of offences against the Law by the ordeal of the 'bitter waters of Jealousy.' The spirit of this ordeal survived till recently amongst the Hovas, who applied it to all suspected cases. The water, harmless in itself, was changed to a deadly poison by an infusion from the kernel of the tangèna fruit, and the accompanying anathemas are an echo of those recorded in Numbers v. 21, 22: 'If thou findest that he has the root of sorcery, kill him instantly, let him die forthwith, tear his flesh, wring his bowels, tear them into pieces, &c.' But, 'if he be innocent let him live quickly, preserve his head without delay; let him rejoice greatly, like one who has drunk cold [harmless] water.'

Bull and calf-worship, common amongst the Israelites and other Semites, also prevailed amongst the Hovas, who made images of bulls, which were bought in the markets, and set up for adoration all over the country.

Among the Jews the New Year was ushered in by the feast of the Passover. So also among the Hovas the New Year begins by a general feast, in which there are certain analogies with the Passover and some other Jewish feasts. Each family group selects a bullock to be killed on the morning of the appointed day. The king or queen also selects a kind of representative bullock to be killed on the morning of the appointed day. The country is searched over for this animal, which must be without blemish and of perfectly symmetrical appearance. A young man, also without blemish or defect, is appointed to immolate the bullock. On the eve of the feast the sovereign after bathing walks about among the people, blesses them and sprinkles them with water from a bullock's horn. Then all feast together, and next morning the bullocks are killed, and some of the blood on a rush fixed to the wall or roof over the doors of all the houses.

Sacrifices on high places were common to Jews and Malagasy alike. Like the Canaanites, the Hovas have sacrificial altars scattered all over the land, whither they bring offerings to various divinities, to obtain long life, health, prosperity, children, and even forgiveness of sins. As the Jewish *Ashera* could be carried from place to place and burned with fire, so the Malagasy had their itinerant idols fixed on slender poles, and also consumed with fire. 'At the coronation of Radama II. it is said he had the whole of them—amounting, as some say, to one hundred more or less, of every kind—brought to take their place in the large assembly. Such a collection of them, if arranged in one place, might be called a grove [*ashera* = 'grove'] by their devotees, and it is now well known that these idols were all afterwards *burned with fire*.'

Of course human nature and its manifestations are very much alike everywhere, and there are startling coincidences, such as the fire-dance, the *couvade*, and the were-wolf—were-tiger, were-panther, were-jaguar, &c.—which seem to girdle the globe. But where we have an accumulation of parallelisms and even identities such as those here specified a common origin will naturally be sought for them. The first impulse will be to turn to the Malay lands, whence came the Malagasy people, but where none of these coincidences are to be found. The only other possible *provenance* is Canaan, where all are found, and whence, as above shown, they could very well have been introduced by the Israelitish crews of David's gold-fleets.

Now these crews were associated a little later on—in Solomon's time—with the Phœnician crews of Hiram's gold-fleets, and we accordingly find in Madagascar some practices still surviving, which appear to be more characteristic of the Phœnician than of the Israelitish Canaanites. Such was that feature of the Phoenician cult which blended the worship of Moloch and the other gods with that of the moon, by most of the Semitic peoples called 'The Queen of Heaven.' This was a highly contagious form of worship, which passed both into Israel and into Madagascar. At times it was all but universal in Judaea, threatening to dethrone Jehovah and bring about a reversion to the cruder forms of proto-Semitic polytheism. It flourished in the streets of Jerusalem, it was flaunted before high heaven on the flat roofs of the houses, it rolled up in murky flames from every city in Israel, from the summits of Sion, of Gerizim and Ebal, and of every 'high place' throughout Palestine. 'For Solomon went after Ashtoreth the goddess of the Zidonians and after Milcom the abomination of the Ammonites. . . . Then did Solomon build one high place for Chemosh, the abomination of Moab, in the hill that is before Jerusalem, and for Molech, the abomination of the children of Ammon' (1 Kings xi.).

So amongst the Malagasy people there was an old custom which was strictly analogous to the offering of incense to the Queen of Heaven amongst the Canaanites. The new year opened with a new moon, when was kept the great feast of *Fandròana*, which is now, I understand, observed at the full moon of the first month Alahamady (see above). This new moon was anxiously awaited and ushered in by much burning of inflammable materials on two successive evenings, the first on behalf of the sovereign, the second on behalf of the people. 'Bunches of dry grass or hay are fixed on long slender poles, kindled, and then carried about, chiefly by young people, with waving and shouting and rejoicing; and this is so common, even in the present day, that the flames, with their ascending smoke, may be counted by hundreds or even thousands.'[73] The spectacle must have presented a striking resemblance to that which was witnessed on every hillside of Canaan at the offering of incense to the Queen of Heaven, the New Moon.

Malagasy oral literature is extremely rich in myths, legends and those other echoes of the past which are generally comprised under the expression *folklore*.[74] Some of these are undoubtedly of local origin: others may be traced to the primeval homes of the natives in the Malay lands, but others again present such close parallelisms with our European popular stories and fairy tales that northern contact and influences at once suggest themselves. I well know what slippery ground folklore is, like South Africa the grave of many reputations. The question of the source of the subjoined instances may, therefore, be left to such recognised experts as Mr. Andrew Lang, Mr. Edward Clodd, or Mr. William Gomme. But if the myths were introduced from the north, then the only possible channel was that of Hiram's Phœnicians, for they alone had relations with all the prehistoric peoples along the shores of the Mediterranean, from whom they may have gathered many of these very ancient popular legends. The Phœnicians maintained constant intercourse with the peoples of the Aegean Sea, where many of our nursery tales were already familiar to the Thracians and proto-Hellenes from remote times. Such was, for instance, the story of Cinderella, the *Aschenputtel* of the Teutons, and the *Ashpitel* of the Scotch, the main features of which reappear in an Eros and Psyche myth preserved by Apuleius (second century A.D.).

I have not met a Cinderella in Madagascar, but there is a 'Beauty and the Beast,' which reads like a variant of its European prototype. The full text occupies no less than forty-eight pages in Mr. Dahle's collection. But the subjoined brief epitome which I prepared for 'Man Past and Present' (p. 256) will suffice for our purpose. 'It is entitled the "History of Andrianarisainaboniamasoboniamanoro," and besides this prince with the long name, called *Bonia* "for short," there is a princess, *Golden Beauty*, both being of miraculous birth, but the latter a cripple and deformed until found and wedded by Bonia. Then she is so transfigured that the *Beast* is captivated and contrives to carry her off. Thereupon follows an extraordinary series of adventures, resulting of course in the rescue of the Golden Beauty by Bonia, when everything ends happily, not only for the two lovers, but for all other people whose wives had also been abducted. These are now restored to their husbands by the hero, who vanquishes and slays the monster in a fierce fight, just as in our nursery tales of knights and dragons.'

There is a legend like that of the Hebrews about 'the sons of God' who 'saw the daughters of men that they were fair,' and another quite after the manner of 'The House that Jack built.' In this all the characters boast of their strength, starting with Ibotity, who climbs a tree, and ending with God, who 'killed men, men maimed crab, crab pierced rock, rock struck boat, boat crossed water, water quenched fire, fire burned iron, iron cut rope, rope caught cat, cat ate rat, rat bored wall, wall stopped wind, wind snapped tree, tree broke Ibotity's leg: *Nothing so strong as God.*' Compare also 'This is the dog that killed the rat, &c.'

But we must not linger over this fascinating theme, else we shall never reach our goal—the golden Havilah, on the opposite side of the Mozambique Channel.

Enough has in any case been brought forward in this and in the preceding chapter to show that both the ancient Himyarites and the Jews and Phœnicians of David's and Solomon's time had established protracted trading and social relations with Madagascar.

CHAPTER XIV

HIMYARITES, JEWS, AND PHŒNICIANS IN RHODESIA—CONCLUSION

FROM Madagascar the transition to the Rhodesian gold-fields could have presented no difficulty, and indeed must have been inevitable once the report was spread abroad that the region west of Sofala Bay was richly mineralised. Cameron (*loc. cit.* p. 9) aptly points out that the joint fleets of Hiram and Solomon undoubtedly passed the island of Socotra which lay a little to the west of our Ophir, on the south coast of Arabia. 'The voyage thence to the east coast of Africa, to the Comoro Islands, or even to the west coast of Madagascar, would be just such a voyage as the Phœnicians in their ships could manage, perhaps quite as well as the Arabs of the present day. Indeed, after accomplishing, as we know they did, the voyage from Ezion-geber to near Socotra, another voyage of about the same length, but much less dangerous, would bring the ships of Hiram and Solomon to Bembatoka Bay which receives the Betsiboka, the chief river on the west coast of Madagascar; while much shorter and not more dangerous would be the voyage from Bembatoka Bay to the bay of Sofala.'

Now we can take the final step on our long excursion, and show that this last voyage across the Mozambique Channel, nowhere much more than 250 miles wide and in many places studded with islands, was actually made both by the Himyarites and Phœnicians. Indeed, this point scarcely needs further discussion after the statements made at the opening of the enquiry, and those more particularly which have reference to the recent explorations of Dr. H. Schlichter, Mr. Franklin White, and especially Messrs. Hall and Neal. There is one factor, however, which, taken in connection with the whole body of evidence, may be regarded as conclusive. About the Phœnicians scarcely any doubt remains. But it might still be asked, were they preceded by the Himyarites, this being one of my main contentions? I have above shown that the more important Semitic elements in Malagasy are derived not from Koranic Arabic, but from the far more ancient Minaean and Sabaean of the South Arabian rock inscriptions. To complete the argument it will suffice to show that an actual fragment of such inscriptions was found by Bent at Zimbabye. Speaking of three shards of three very large bowls, all presenting exceptional interest, Bent writes: 'The third fragment is perhaps the most tantalising of all; it is a fragment of the lip of another large bowl which must have been more than two feet in diameter, and around which apparently an inscription ran. The lettering is provokingly fragmentary, but still there can be no doubt that it is an attempt at writing in some form: the straight lines down the middle, the sloping lines on either side recall some system of tally, and the straightness of the lettering compares curiously with the proto-Arabian type of lettering used in the earlier Sabaean inscriptions, specimens of which I here give, and also with some curious rock carvings found by Mr. A. A. Anderson in Bechuanaland' ('Ruined Cities,' p. 167).

The Bechuanaland 'rock carvings' are mere scrawls, though to the trained eye they certainly do present a curious Himyaritic *facies*. Very different are the Zimbabye characters which are compared by Bent with a number of proto-Arabian letters rightly described by him as 'used in the earlier Sabaean inscriptions.' The 'earlier Sabaean,' as we have seen, was Minaean, and the four or perhaps five letters here reproduced from Bent find their exact counterpart in the oldest Minaean

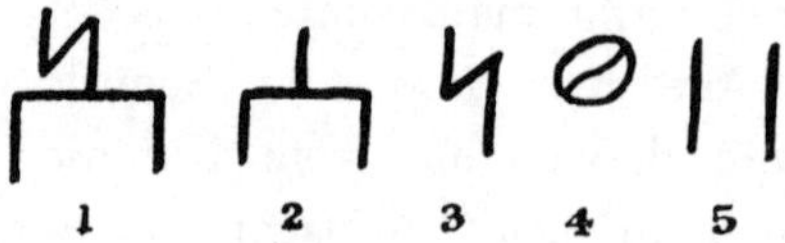

carvings. They are in fact quite legible, those which I have numbered 1, 2, 3, 4 being equivalent to our *a*, *s* (thick), *n* and *w* respectively. They all occur repeatedly and with great uniformity; for instance, on the inscribed stone which is usually quoted as *Glaser* 282, being amongst the spoils captured by Herr Eduard Glaser in 1885, and presented by him to the British Museum. The fifth is doubtful, and may be either a *b*, where the horizontal top line has been obliterated, or else simply two vertical dividing lines, such as are common in

Himyaritic documents. Either way we have an undoubted fragment of the Minaean script, in which was presumably written the now lost inscription which was still conspicuously shown on the Zimbabye monument four hundred years ago, and has already been referred to.

There is another such fragment, which was discovered by Dr. Schlichter in one of the numerous slave-pits of the Inyanga district, and figured by him in the 'Anthropological Journal,' April 1899, p. 384. In this remarkable district, one of the most densely peopled in Himyaritic times, 'there are hundreds of ancient forts and extensive citadels, the last-named being used for defensive purposes against the surrounding mountain tribes. Indications of the ancient Semitic stone and solar worship are numerous, and in one of the slave-pits of Inyanga I havė discovered on an old soapstone slab *the remnants of an inscription, which is undoubtedly* of ancient Semitic origin . . . Several letters [five are shown] are still perfectly legible and well incised. It has an archaic and quaint appearance [not more so than many from South Arabia], and I have already mentioned that it is of ancient Semitic origin' (p. 383). I do not think there can be the shadow of a doubt on this point, all the letters, which I transliterate *z*, *k*, *a'*, *ṣ*, *d*, being quite clearly drawn, and, as far as one can judge, resembling the oldest known Phœnician more than the ordinary Himyaritic forms. They also seem to run horizontally from right to left, as usually with this script.

I do not know whether any other graven documents have since come to light beyond two stones with (probably)

Himyaritic inscriptions which Dr. Peters writes me he found during his last expedition, but could not then remove, and his attendant was afterwards unable to recover owing to the tall grass ('Letter,' Oct. 17, 1901). But explorers should be on the look-out for such relics, and in order to facilitate research I have now reproduced in facsimile the above quoted Glaser document *in extenso* (see frontispiece) from the 'Babylonian and Oriental Record' for September 1887. The inscription occurs on a stone discovered with hundreds of others by Glaser during his second journey to Yemen. The stone, which came from the Jauf district, a little east of Maraiaba, was in five fragments, but has been pieced together by the finder himself, so that the sequence of the text can be trusted.

In our reproduction the lines are disposed in the direction from right to left in order to facilitate comparison with the other Semitic scripts, all of which also run from right to left, the Abyssinian group (Geez, Tigraic, Amharic) alone excepted. The Himyaritic rock inscriptions themselves, as shown in Fritz Hommel's 'Süd-Arabische Chrestomathie,' Munich, 1893, run in various directions, some being even disposed in vertical columns, like the lettering sometimes seen running down the sides of our shop-windows. This should be carefully noticed by explorers, who may have to turn inscribed blocks one way or another to detect any resemblance with our facsimile.

One of the most striking features of this very ancient writing system is its remarkable uniformity. Not only does each letter occur again and again nearly always in the same form and size, so that it can everywhere be recognised, but each word appears to be invariably marked off from the preceding and following by a straight line. An analogous arrangement prevails in the numerous Axumite inscriptions of Abyssinia, but the diacritical devices are different and of quite a unique character. I have no reason, however, to suppose that any of these have ever found their way to South Africa.

It may be mentioned that the Sabaeo-Minaean script is purely alphabetical, like the Hebrew, the Phœnician, and all its European offshoots, whereas the Geez and its modern representatives are syllabic like Sanskrit: *ka*, *ke*, *ki*, *ko*, *ku*, *kau*, &c. Should any of the Axumites have strayed into Rhodesia, traces of this peculiar orthographic system may possibly also turn up.[75]

I need not here repeat the general account of the ruins and workings given in Chapter II. But I will ask the reader to go back to p. 17, where is given the classification of the remains according to Messrs. Hall and Neal. Nobody has a better right to speak on this delicate subject, and when they tell us in Chapter XII. of their book that the scheme is an induction from the actual survey of some two hundred sites, it will be admitted that it has at least a tolerably substantial foundation. In any case, the first two classes—those with which we are most concerned—are established beyond any reasonable doubt. Not only is the treatment different, but the difference is such as to imply a considerable lapse of time between the two orders, while the second is often clearly seen to be either an extension or enlargement of the first, or even actually superimposed on its foundations. 'The second period architecture is also to be found in the shape of most obvious extensions of, and additions to, the first period buildings in the area in which the original buildings are only those of the first period [that is to say, in those districts where all the original structures belong exclusively to the first period], and these additions and extensions are generally built over and upon the original erections' (*op. cit.* p. 165). Thus the time sequence becomes as obvious as we see it, for instance, in the early Norman and later English sections of Canterbury Cathedral.

Again: 'Every archæologist who has written on the subject emphatically asserts that the ruins seen by them and which they mention by name show several, or at least two periods, if not three periods of architecture and construction; and the terms *best period, original period, later period of ancient building, decadent period*, and similar terms to describe differences of architecture and construction are very frequently employed by them' (p. 165).

The more striking points of difference between the first and second styles, that is, as we shall see, the Himyaritic and the Phœnician, will be best understood by the subjoined summary of each in parallel columns:

First Period (*Himyaritic*)	Second Period (*Phœnician*)
Massive strength; enormous solidity; monotonous plainness and	Construction generally less massive and solid; main walls

FIRST PERIOD·

(*Himyaritic*)

uniformity; perfect symmetry; walls thick enough for a 16-team ox-waggon on their present summits with room to spare; elliptical forms dominant, with a system of curves revealing great geometrical knowledge, and an orientation suggesting extensive astronomic lore, especially that of the northern hemisphere and zodiacal science; solstitial positions indicating a period of from 1100 to 2000 B.C.; planned primarily for protection, with difficult approaches, narrow, almost funnel-shaped entrances buttressed for defence, and labyrinthine passages also protected by buttresses and traverses; main and divisional entrances and ends of walls always rounded off; foundations of main walls always carried down to the bed rock; batter-back at least one foot in six feet both inside and outside of main walls, less on parti-walls; no rectangular main walls, but divisional walls disposed at all angles from the inner side of the main walls; no false courses; the walls bonded throughout their whole width, the internal stones being carefully laid; outside and sometimes inside mural faces diversely ornamented; surfaces always smooth and even; courses regular, each stone closely fitting into the next, and the joints always overlapped by a stone above them; floors laid with the finest granite-powder; cement always beautifully smoothed and glazed; cones and conical buttresses with plat-

SECOND PERIOD

(*Phœnician*)

much slighter, mostly no more than three feet thick at their base. Terraced walls are characteristic of this period; they are built up with receding ledges against the slopes, the foundations being usually laid on rough granite blocks, which are seldom more than from eighteen inches to two feet from front to back; orientation appears to be irregular, lacking system or uniformity, hence possessing no astronomic significance; main and divisional entrances and ends of walls squared and not rounded; are comparatively plumb; approaches easy, the main entrances opening directly into the interior of the building, and evidently erected in more peaceful times when defensive works were less needed, as if the aborigines had now been completely crushed; foundations of main walls often not carried down to the bed rock, though it may be no more than two or three feet farther down; retaining walls often rest on the soil; less batter-back on outside of main walls, none inside; right angles in main walls prominent and general; false courses frequent, and interior of the walls carelessly built; stones between the two faces thrown in anyhow; a greater profusion of decorative work on the outer face of main walls, and on the faces of the terraces most elaborately carried out; check pattern most prevalent; in general the masonry, though on the whole good, shows symptoms

First Period (*Himyaritic*)	Second Period (*Phœnician*)
forms on their summits always well finished; drains general; steps rare; gold ornaments found on the original floors massive, quite solid, well executed, and abundant; much of the gold extracted from the mines evidently worked up on the spot for the local demand; gold-smelting furnaces numerous; no copper or iron ornaments; site often low knolls rising above the plateau. 'The exact similarity of the ruins of the first period in architectural features to those of the Sabaeo-Arabian temples is very striking, and points to the earliest Zimbabwes having been erected by the Himyaritic or Sabaeo-Arabian people' (*op. cit.* p. 160).	of falling off; the inside walls being very inferior and without decoration; the granite-powder cement of the floors began already to deteriorate both in quality, thickness, and finish; no cones or conical buttresses with platforms on their summits occur anywhere; drains absent; steps of granite blocks cemented over a common feature; the gold ornaments far less abundant, inferior also in make and value; thus copper and even iron bangles occur with gold bands or beads at intervals; no gold-smelting furnaces found anywhere, showing that most of the gold was now won for exportation, probably in the form of gold dust, this being due no doubt to the increased value of the gold standard in the markets of the world; site mostly the summits of high kopjes, except of course where the later structures stand on the same sites as the earlier, with which they are so often intermingled. Many observers think the monuments of the second period show greater affinities with Phœnician than with Himyaritic prototypes. But the distinction is not always quite obvious in the Semitic lands themselves.

A general characteristic of both groups was the absence of roofed buildings. 'In not one of over two hundred ruins in Rhodesia which have been examined, are there any evidences that the buildings, or even the enclosures, were ever roofed. Indeed, the extent of some of the enclosures, and their most irregular shapes, entirely preclude all probability of roofing being erected. Every archæologist who has inspected the ruins has arrived at the same conclusion. Certainly the temples in which Phallic worship was conducted were without roofs, for, in the old historic words of writers on this subject, these were all "open to the light of heaven"' (*ib.* p. 183). So true is this, that the obverse of the Byblos coin, above alluded to, seems singularly like a reproduction in miniature of some of the ground plans of the Rhodesian structures.

The close resemblance to the South Arabian method of construction, above referred to, is seen, especially in the small, brick-shaped, granite blocks, laid together evenly, but without mortar: 'The large, circular building of the Great Zimbabwe ruins is built of small blocks of granite, broken with the hammer into a uniform size, but bearing no trace of chisel marks, and no mortar had been used in the construction. The same kind of construction, viz. hewn stones neatly put together without mortar, was found by Halévy in the ancient and great ruins of the city of Me'in in South Arabia, only that in this case the stones are larger. Halévy found the ruins of Me'in to be in all probability the ancient capital of the Minaei, the *gens magna*, a tribe about the position of which, in the modern map of Arabia, geographers for a long time greatly differed.'[76]

Returning for a moment to the two periods, it is important to note that, except where intermingled, they occupy two different geographical areas. Here the surveys of Messrs. Hall and Neal, with their associate Mr. Johnson, are of great value for determining the sequences, and forming a clearer view of the relations existing between the two or three successive waves of Semitic immigration. 'The first period buildings are found on the east, south-east, and south of a line drawn from the north of Zimbabwe to the north-west of the Matoppas, with the exception of a few ruins overlapping on the western side of this line.' But those of the second period occur 'to the north, north-west, and west of this line, extending from and including Inyanga . . . down in a south-westerly direction to the western border of Bulalema district, south-west of Bulawayo.'

Thus, as might perhaps be expected, the area occupied by the earlier remains lies more to the east, that is, nearer to the coast, and this region would naturally be the first to be reached and settled by the pioneer intruders. Why, then, it might be asked, did not their successors also occupy the same district? The reply is that they did, as shown by the later extensive additions to the original buildings; but that, according as the eastern workings became less profitable, they spread gradually westwards, until a great part both of the Mashona and Matabili lands was occupied.

I think it may now be shown that the first arrivals were the South Arabians, the second the Phœnicians and their Jewish associates. There is now in the possession of Mr. Cecil Rhodes the most precious relic of bygone times hitherto found south of the Zambesi. This is the famous 'Zimbabye Zodiac,' a large offering-dish of extremely hard wood, picked up some miles from Great Zimbabye, and showing the twelve signs of the Zodiac carved round the rim, just like the eighteen month-signs on King Axayacatl's great calendar-stone, now in the Mexican National Museum. Dr. Schlichter, who has carefully examined and reproduced it in the 'Geographical Journal' for April 1899, p. 387, tells us that the signs 'coincide in every respect with other finds which Bent and others have made in Zimbabye. One of the pictures is an image of the sun, analogous to the sun-pictures which Mauch and Bent found on the monoliths of Zimbabye, and *analogous also to finds in Asia Minor which belong to the Assyro-Babylonian* period. And this image of the sun has its position directly behind the Zodiacal image of Taurus, and between the signs of Taurus and Gemini. But we know that in early antiquity the beginning of the year was represented by the Zodiacal sign of Taurus. This was so firmly established (although not correct any longer at the later time of the Roman period) that even Vergil still sings—

> Candidus auratis aperit cum cornibus annum
> *Taurus.*

Therefore, this position of the image of the sun on our Zimbabye Zodiac points directly to the early times of antiquity' (*ib.* p. 382).

And again: 'We have in Zimbabye an enormous gnomon before us, comprising a total angle of 120°. Taking all the details into consideration, I find that the obliquity of the ecliptic was somewhat more than 23° 52', which brings us to a time somewhat 1100 B.C. for the erection of the Zimbabye ruins' (*ib.* p. 387).

This date, determined astronomically, shows that the first arrivals in Rhodesia and the builders of Zimbabye must have been the South Arabians, and not the Jews and Phœnicians, who did not begin to despatch their joint expeditions from Ezion-geber till the time of Solomon, that is, fully a hundred years after the erection of the 'gnomon.' This harmonises completely with all the other conditions, especially the superior workmanship of the early monuments, corresponding with the now ascertained superior culture of the Minaean and Sabaean nations, and also with the already mentioned presence of Himyaritic characters on the oldest structures. If this does not prove the early advent and long sojourn of the South Arabians in Rhodesia, then I will ask, How do the analogous monuments and inscriptions prove their presence and long sojourn in Abyssinia, where nobody doubts that they were the founders of the Axumite empire? I claim, therefore, that this part of the thesis is established irrefragably.

Coming now to the Zimbabye Zodiac, is it not equally self-evident that this document records a calendric system constructed at a time when the year began at the vernal equinox with the entrance of the sun into the constellation Taurus? Since then, owing to the precession of the equinoxes, Aries has gradually moved into the space assigned by the ancient astronomers to Taurus, and has consequently become the *first* sign of the Zodiac. Moreover the shift took place at such an early date that Aries itself has, so to say, nearly run its course, and the first point of this constellation has already moved well into Pisces.

We have therefore to go back some millenniums to reach the period when the sun entered Taurus on March 20. So remote in fact was this period that Taurus became the first sign of the Zodiacal year amongst all the nations of antiquity that borrowed their calendric systems from the Assyro-Babylonian astronomers, the reputed constructors of our Zodiac. Not only so, but the very word *Taurus* itself accompanied the system in all its peregrinations amongst the surrounding populations from Mesopotamia, or possibly South Arabia eastwards to India, southwards to Madagascar, and westwards to the shores of the Atlantic. I say possibly South Arabia because since the disclosures made by the interpreters of the Himyaritic inscriptions it has become a moot question whether Arabia Felix rather than Babylonia may not have been the cradle of human culture.

But however this be, the word Taurus would appear to have been transmitted from the Semites to the Aryans before the Aryan dispersion itself. Philologists are of accord that a term found in all the great branches of the Aryan linguistic family must have belonged to the common stock of the primitive Aryan mother tongue before it broke up and sent its ramifications over a great part of the northern hemisphere. Such a word is Taurus, as may be seen from the subjoined comparative table, showing its immense range throughout the Semitic and Aryan lands, together with the Malagasy section of the Indo-Polynesian world :—

Semitic Tongues		Aryan Tongues.	
Chaldaean	thor	*Zend*	staora
Hebrew	shor	*Greek*	tauros
Syriac	thauro	*Latin*	taurus
		Italian	toro
Himyaritic	tsauru	*Gothic*	stiur
Arabic	tsauro	*Old German*	stior
Arab. Variant	tauru	*Anglo-Saxon*	steor
Arab. Dialect	toro	*English*	steer
		Old Slav	turu
Malagasy (Hova)	adaoro	*Norse*	thjorr
,, (Fort Dauphin District)	tzoro	*Swedish*	tjur
		Danish	tyr

Here the prosthetic *s* of some Aryan forms seems best explained by the South Arabian *ts*, although a Teutonic root *steur* referable to the Sanskrit *sthura*, huge, mighty, has been postulated. This universality of Taurus, ranging literally from Madagascar to Iceland, is the more remarkable, since it stands quite alone, and is not shared by any of the other Semitic names of the Zodiacal signs. Some might perhaps say that an exception should be made in favour of Aries, the *Al-hamalu* of the Arabs, with here also the Malagasy variants *Alahamady* and *Alahamalu*. Dahle thinks this may be the same word as the German *Hammel*, a wether (*loc. cit.* p. 78). But it cannot be so, and the resemblance, though curious, is merely a coincidence. Hammel has nothing to do with Semitic, being distinctly referable to a Teutonic root *ham*, to maim or cripple, which still survives in the Anglo-Saxon *hamelian*, and the English *hamble*, to mutilate, &c. For the *b* compare *humble* from the Latin *humilis*, whence *humility* without the *b*.

More important is the question whether the Semitic root itself may not have been derived from the Aryan, or both from a common primeval Semito-Aryan tongue. Such a tongue is unknown to science and almost unthinkable. Every attempt made to reduce the two groups to a common stock has not only resulted in failure, but has tended to widen the breach between them.

That the Aryans were here the borrowers, the Semites the givers, is evident from the fact that the latter were the inventors of the Zodiacal system, and that at a time when the vernal equinox coincided with the entrance of Sol into Taurus. Hence the pre-eminence of this constellation, which, as shown by the above quotation from Vergil, survived even its displacement by Aries. Vergil was wrong when he sang that 'the white bull opened the year with his gold-tipped horns,' for at that time his place had already been usurped by Aries; only the tradition was still too strong for Taurus to be ousted except by science, always in the van.

Now this Zodiacal system was at first concerned more with human destinies than with celestial measurements.[77] As alchemy preceded chemistry, so astrology led the way to astronomy. Hence the planets and their conjunctions and right ascensions, and all the rest of the perverted jargon, were manipulated, not by the man of science, but by the priest, the magician, the soothsayer, the wizard, the *Zekuru*. This word Zekuru is Chaldaean, that is the Semitic dialect of Southern Babylonia, and from the Chaldaean it passed through the Arab *sikhr* to the Malagasy *sikiry*, with its variants *sikily*, *sikidy* (see above), and through the Syriac *Zakuro*, a soothsayer, to the Greek *ζάκορος*, a priest, or rather a minister, a keeper of a temple (*ædituus*, *famulus*, &c.), as seen in its one derivative *ζακορεύω*, to ministrate. The Greek is thus a rare oriental exotic pointing to Babylonian influences, and again proving that the Zodiacal terminology is not Aryan, but Semitic. Its survival in Madagascar must consequently be due to the same influences, and we may confidently conclude that the priests and wizards, who presided at the religious observances of ancient Rhodesia, and interpreted the 'Zimbabye Zodiac' for their votaries, were the South Arabian *sikhr* (Zekuru), who have left their very name in the neighbouring island of Madagascar.

Did they leave their very bones also? Some have already been found, which may plausibly be assigned to them (see above). It was certainly a surprise and disappointment to the upholders of the Semitic theory when Bent announced that he had nowhere discovered any graves or burial-grounds of the Arabian immigrants. It was naturally inferred that they could not have been permanent settlers in the country, that their sojourn could not have been protracted and continuous, despite the evidence of the monuments. The difficulty was not overcome by the suggestion of cremation, because those Semites, as is well known, did not burn but buried their dead. Many parts of Canaan are strewn with remains of a dolmenic type, and we know that dolmens were in their inception the tombs of the departed, especially chiefs and 'druids.' But the difficulty was got over, as might almost have been anticipated, in the most convincing manner, by the discovery of the graves and cemeteries themselves. 'I have found,' writes Dr. Schlichter (*loc. cit.* p. 386), 'north of Inyanga [beyond question one of the most thickly peopled settlements of the foreigners], what I believe to be an *extensive burial-ground.* Owing to the absolute want of boys [native assistants] at the time, I was unfortunately prevented from doing any excavation work thereon. It is a great pity that when, some time back, some prospectors of what is called "The Ancient Ruins Company" found *a number of ancient graves*, and got out of them considerable quantities of gold, they did not keep the human skeletons which they found. Although I took the greatest trouble to ascertain details, I can only state, as far as my information goes, that the *skulls were dolichocephalous.*'

This also is what might have been anticipated, since the Semites are distinctly a dolichocephalous (long-headed) race, with cephalic index as low as 72°. The same might no doubt be said of the Hottentots and Bushmen, whose average index is also about 72° or 73°. But the remains in question were not those of the Hottentots, who were the slaves employed in the mines, and slaves are not buried with 'considerable quantities of gold.' [78] The remains were obviously those of their taskmasters, who intermingled little, if at all, with these low-type natives. In 'Man Past and Present' (p. 88) I remark that 'no doubt other Semites (Minaeans, Sabaeans, and Himyarites generally) almost certainly [I should now say certainly] reached the east coast below the equator in early historic times. But they appear to have arrived chiefly as traders and miners, and never to have penetrated far inland, except in the auriferous regions south of the Zambesi, where their still extant monuments in the Zimbabye and other districts show that they held the country by military tenure, and mixed but slightly with the Negro aborigines.'

Nevertheless distinct traces of Semitic interminglings with the present Negroid populations are stated on good authority to be everywhere perceptible in the region between the Lower Zambesi and the Limpopo. Bent's testimony is strong on this point, which certainly calls for more elucidation than it has hitherto received. At p. 62 of 'Ruined Cities' is figured an impressive group of high-caste natives—Umgabe, and his Indunas [chiefs]—of whom Bent writes: 'Umgabe is a huge fat man, tall and dignified, though naked; around his neck he has a string of large white Venetian beads of considerable antiquity, brought doubtless to this country by Arabian traders in the Middle Ages; in his hand he carries his iron sceptre, the badge of a chief, and his battle-axe is lavishly decorated with brass wire. Amongst his men we saw many of varied types, some distinctly Arabian in feature, and I am bound to say the Kaffir type amongst them was the exception and by no means the rule. Arched noses, thin lips, and a generally refined type of countenance are not, as a rule, prominent features amongst those of pure Kaffir blood, but they are common enough around Zimbabwe.'

In fact, we have before us a group of exceptionally fine Bantus, and the ethnical relations south of the

Zambesi will never be understood until we thoroughly realise the meaning of this term 'Bantu.' On this subject I elsewhere write ('The Boer States, Land and People' p. 96 *sq.*): 'That the streams of Bantu migration had their source in the East Equatorial lands, and set along the eastern seaboard continuously southwards to the extremity of the continent, is shown amongst other grounds by the term *Munkulunkulu*, one of the Bantu names of the Deity, which may be followed in its numerous variant forms along the whole route from the Tana basin near the Equator to the Great Fish River, the southern limit of their domain in Cape Colony. Thus:—

Name of Deity	Tribe	District
Mulungu .	. *Wanyika*	Coast north of Mombasa
Mungo .	. *Wapokomo*	Banks of the Tana
Mulungu .	. *Wagiryama*	Coast between Melinda and Mombasa
Muungu .	. *Waswahili*	Zanzibar coast
Mulungu .	. *Waruandi*	East side of Lake Tanganyika
Mungu .	. *Walunda*	Lake Bangweolo
Mulungu .	. *Manganja*	Lake Nyasa
Muunga .	. *Makua*	Mozambique
Murungu . (*r* = *l*)	. *Batoka*	Manica and Sofala
Unkulunkulu	. *Zulus*, *Xosas*	Zululand, Kaffraria
Ukulunkulu	. *Amampondo*	Pondoland

The term *Bantu* is a native word meaning 'people' (from *Aba* or *Ba*, plural form of *um*, *umu*, the personal prefix, and *ntu*, a man, a person). It was first applied by Dr. Bleek to all the peoples south of the Zambesi who are neither Bushmen nor Hottentots, but speak various forms of a now extinct stock language, and are therefore assumed to belong to one common racial group. As they never had a common ethnical name, they were all conveniently comprised under this general designation of Bantu, by which is therefore to be understood 'peoples of Bantu stock and speech.'

Later, the discovery was made that the Bantu linguistic family had an immense range extending, in fact, almost exclusively over the whole of the continent from Kaffraria to about four or five degrees of latitude north of the Equator.

In this vast domain of some six million square miles, no other languages are known to exist except the Hottentot, Bushman (with Sandawi), a few Negro tongues penetrating south from Sudan, and the non-Bantu idioms of the Negritos dispersed over the Congo forest zone. Thus it is that we now hear of Bantu peoples and Bantu languages everywhere. But the term is still somewhat restricted to scientific writings, and has not yet obtained currency in the British colonies, and south of the Zambesi, where the popular collective name of these natives is Kafir or *Kaffre*.

Here it is carefully to be noted that the Bantus are not full-blood Negroes like those of Upper Guinea and Sudan, but a *Negroid* people, that is, a mixture of two or more elements, of which the substratum is the Negro, and the later infusions mainly Hamitic (*Galla*), and, to a less degree, Semitic (*Arab*). The mixtures took place at different times and in varying proportions, the result being that the Bantus themselves show nearly all shades of physical and mental characters, intermediate between the pure Negro or Ethiopic and the much higher Hamitic and Semitic (Caucasic) types of mankind. Hence the extraordinary differences that are observed between, for instance, the degraded Magwamba ('demons' or 'devils'), called Knobnoses by the Transvaal Boers, and the Basutos, Zulus, and others, who are both mentally and bodily immensely superior to the Felups and other true Negroes of Senegambia.

Everything now seems clear. It was above shown (p. 28) that there were no Bantus but only Hottentots and Bushmen in the Sabaeo-Phœnician period, when the monuments were built and the mines opened. Then came the Bantus, fierce, warlike peoples descending along the east side of the continent, and sweeping everything before them. It is to the irruption of these Bantu hordes into Rhodesia that is to be attributed the sudden suspension of the mining operations, the expulsion or extermination of the Semitic prospectors and settlers, and the rejection of the Hottentot aborigines to the south-west corner of the continent. At the Mundie ruins we saw (p. 19) that there were found unmistakable indications of conflicts, massacres of the old occupants —unburied remains, torn gold-wire bangles, scattered beads, and so on. There was unquestionably a total interruption of the works, and of the traffic in gold, until these relations were afterwards renewed under different conditions by the 'Moors,' that is, the Moslem Arabs, whom the Portuguese found trafficking in gold with the natives at Sofala and in Manicaland (*supra*, p. 1).

But Mr. Selous holds that there was no interruption and that true mining processes, quartz crushing, and not merely alluvial washings as described by De Barros (p. 4) were continued by the natives themselves down to quite recent times. This is a most interesting point, on which

Mr. Selous is entitled to have his case stated in his own words. In the Geographical Journal,' April 1899, pp. 391–2, he says: 'The two principal arts known to the ancient people appear to have been the art of building walls of stone bricks carefully fitted together without mortar, and the art of extracting gold from quartz reefs, and I think it is generally considered that these two arts died out in South-Eastern Africa a very long time ago, at the time when the ancient people who first introduced them into the country are supposed to have been swept off the face of the earth in some catastrophe—possibly a rising of the aboriginal races, during which they and their civilisation died together. But I believe that there is an intimate connection between the Bantu tribes now inhabiting South-Eastern Africa and the ancient Semitic people who first built temples and extracted gold from quartz in that country. Not only did the extraction of gold from quartz not become a lost art a very long time ago, but we have overwhelming evidence that it has been continually practised down to quite recent times by the immediate ancestors of the present Bantu tribes, and was only discontinued on account of the Zulu invasions into Mashonaland, which have taken place during the present century. Old men who accompanied Umziligazi [Lobengula's father] on his first incursion into the country we now call Rhodesia about sixty years ago [1838?] have repeatedly told me that they found the *Amaholi*, as they termed the native inhabitants of the country, working in the *amaguti*—that is, in the deep holes—the old workings which have recently been found by Europeans on quartz reefs all over the country. In

1891, too, Mr. Kock, a prospector, found a bucket and rope made of *machabel* bark at the bottom of a shaft 120 feet deep on a reef near the Umzweswi river. Articles made of such perishable materials could not have been very old, and they had obviously been used to haul up quartz from the bottom of the shaft. In 1884, too, I visited with Mr. S. H. Edwards the mouth of an old shaft filled up with *débris* in western Matabeleland. 'By the side of the shaft was a heap of quartz, which had evidently been burnt and stacked ready for crushing, and by the side of the heap of quartz lay several of the hard round stones with which the natives used to crush it before extracting the gold. The burnt quartz, after having been broken into little pieces, was laid little by little on a large flat stone, and ground fine with a large round stone. The gold was then washed out of it and melted into small ingots in clay crucibles. On opening up the shaft I have just spoken of we found that in one part the roof had been supported by stout poles of *mapani* wood (the common wood of the country). These poles I examined carefully, and found that they all had their bark on, and showed no sign of great age, and had, moreover, all been chopped with the same kind of narrow-bladed axes which the natives use at the present time. In my own mind I feel sure that work at this mine, and at others in the same district, only ceased about sixty years ago, when the Matabele first invaded the country [after their defeat by the Transvaal Vor-trekkers]. But the extraction of gold from quartz was an industry carried on in eastern Mashonaland at a much later date than that.

The testimony of Mr. Thomas Baines, the well-known traveller and artist, is incontestable on the point. He found the Mashonas extracting gold from hard quartz reefs in the neighbourhood of Lo Mangondi's (about seventy miles north-west of the present town of Salisbury) as late as 1870. You will find a full account of his experience in Mr. Baines's book, "The Gold Regions of South-East Africa." . . . Thus you will see that the art of extracting gold from quartz, which was first introduced into South Africa by a people of Semitic race—as Dr. Schlichter considers—some three thousand years ago, has endured in that country until quite recent times.'

Now all the facts here specified are absolutely correct. Of this I have independent evidence, and in any case it would be more than rash to question any statements made by Mr. Selous regarding recent relations in almost any part of South Africa. Yet his inferences I believe to be wrong. Apart from the proofs of sudden interruption, massacre and ruin above indicated, we have De Barros' positive assurance that four centuries ago the Arabs had no knowledge of the old monuments and workings, and that on the advent of the Portuguese the operations of the natives were limited to washings, that 'they dive down and find much gold in the mud brought up,' and 'generally nobody digs more than six or seven spans deep, and if they go to twenty they come upon the hard rock.' How then are they now, or were lately, skilled quartz crushers? How reconcile Selous and De Barros?

Dr. Schlichter gives the clue. After concluding

(*loc. cit.* p. 390) that 'the territories known to-day as Southern Rhodesia were 1,000 years before the commencement of the Christian era a gold-producing country of a large extent, and *colonised by the early Semitic races round the Red Sea, viz. by Jews, Phœnicians, and western Arabians*, he goes on to speak of three old Portuguese fortifications, 'which I found in the northern part of Inyanga ; of one of them I made a detailed plan and measurements. The illustration [figured at p. 385] of the section of this fortification shows that these constructions are totally different from the old Semitic ruins, as well as from the Monomotapa Kafir constructions. The existence of these Portuguese forts proves the correctness of the old Portuguese writers, of the sixteenth and seventeenth centuries, who asserted that the early Portuguese dominions extended far into the interior of South Africa, south of the Zambesi.' It is, moreover, a fact of common knowledge that the Portuguese have always held a firm grasp of the rich auriferous Manica district, where they have carried on mining operations almost continuously for over three hundred years, down to the present time. But they did not work the mines themselves, but, like their Semitic precursors, employed native labour for the purpose. We do the same now in the Kimberley diamond fields and in the Johannesburg 'Rand.'

Thus it came about that the Bantu gold-washers of four centuries ago gradually became the expert quartz-crushers that Mr. Selous and others have found them to be in recent times. There is no contradiction, and the statements of De Barros and Selous are in accord. Only the continuity theory must be given up, and this even on other grounds. By the 'Monomotapa Kafir constructions' Schlichter means those erected by the Kafirs (natives) at the time when the Monomotapa was 'Lord of the Mines' *de facto* as well as in name (see above). Such structures would therefore correspond with those of Hall and Neal's fourth period 'when local races endeavoured, with ill-success, to adopt the Zimbabye style of architecture' (p. 17 *supra*). Here is their description of these 'monuments:' 'Small circular stone buildings, smaller inclosures often made of blocks taken from the more ancient walls, but altogether a poor imitation of the workmanship shown in the earlier periods. The cement is also of a very much coarser quality. These buildings have been very frequently erected on the cemented floors laid over the filled-in inclosures of first and second period ruins, which filling-in process is believed to have been the work of the third or fourth period, probably that of the bastard races left in the country on the departure or wiping out of the ancients; or of local races, upon whom the impression of the ancient methods of building still remained.'

For obvious reasons I cannot accept the latter view. But the general description points not merely at decadence, that is a gradual decline in artistic excellence, without an absolute break of continuity. It points at a reversion to barbarism, a re-occupation of the buildings after an unknown lapse of time by rude tribes, who utilised the materials at hand for their own primitive structures. It seems strange that Mr. Selous should associate such buildings with the highly finished and scientifically planned monuments of the Himyaritic period.

Besides Dr. Schlichter, who has elsewhere ('Geographical Journal,' July 1893) gone into this very question of continuity, shows plainly that there was a complete arrest of the flourishing traffic in the precious metals for several centuries before and after the new era, and consequently that for a very long time the old workings and buildings had been abandoned. At least for many hundred years there was no export trade in gold from the east coast of Africa, and as we know from the later Arabs (*supra*, p. 5) that the natives worked lazily and reluctantly even at the alluvial washings in their time, when some stimulus was given to the local trade, we may be quite sure that the industry was entirely suspended in the absence of such stimulus during the previous centuries between the overthrow of the Sabaean power and the revival of commercial enterprise by the Moslem Arabs. 'These ruins of Mashonaland were doubtless in connection with the gold production carried on in the surrounding country, and enormous quantities of gold must have been exported from South-Eastern Africa to Arabia at the time when this gold production flourished. Now, I have carefully looked through the statements of the classical writers, but I am unable to discover that gold formed an important trade article of South Arabia during the classical period of antiquity. Of course different authors mention that gold was found in various places of the Arabian peninsula; *but this gold is of subordinate importance.* Herodotus states that the Arabians paid a tribute of 1,000 *talents* of *frankincense* to Darius. Pliny says that the Sabaeans are very rich, because the Arabs always sell, but never buy anything, and according to Strabo this was one of the principal reasons which influenced the avaricious Emperor Augustus to undertake the expedition of Aelius Gallus against the Southern Arabs [he penetrated to their capital, Maraiaba, but had to return *re infecta*]. But nowhere is the slightest hint to be found that gold was a trading article of the Southern Arabs during the classical period . . . Later on, at the time of the Periplus, gold was entirely unknown as a product of East Africa' (*loc. cit.* p. 50). But the Periplus brings events down to at least the middle of the second century, a disastrous time for the Himyarites, who then lost their command of the sea, and never renewed their relations with the long abandoned gold-yielding colonies, although a portion of the east coast as far as Azania (Zanzibar) was still politically dependent on the Sabaean empire.

It may be gathered, I think, from Hall and Neal's general survey, and careful mapping of the whole ground, that the earlier monuments were not only more solidly constructed, but also covered a wider area, and imply altogether a longer residence in the land than their Israelitish and Phœnician successors. This is in full accordance with the historical data as supplied by the sacred texts, from which it might almost be inferred that the South Arabians were the true settlers and colonists, the others both later arrivals and less permanent sojourners. There is no evidence of any Canaanitish expeditions, to use a term which will cover both Israelites and Phœnicians, before the time of David, and although he reigned forty years, it is not to be supposed that he was able to turn his attention to foreign enterprise before he had crushed all his enemies towards the end of his reign, say, at most, ten years before the accession of Solomon. Nor could he despatch fleets before he had a seaport, and he had no seaport before the reduction of Edom, which gave him access to the Red Sea. Hence his somewhat problematical expedition to Urphe (see above) can hardly have, in any case, taken place much before 1020 B.C., for Solomon succeeded in 1015. But Solomon died in 975, and his death was immediately followed by the dismemberment of Israel, and that by the Egyptian King Shishak's capture of Jerusalem in 971. Finally Jehoshaphat's navy was 'broken' at Ezion-geber somewhere about 900, and there was an end of all foreign enterprise on the part of the Jews and their Phœnician associates.

With the death of Hiram, certainly during the long reign of Solomon, for he was already ruling in David's time, the alliance ceased. Nor could the Phœnicians any longer maintain their commercial position against the rivalry of the Himyarites in the Indian Ocean in those troublous times, when Egyptians, Assyrians, and Babylonians were at constant warfare, and Tyre itself besieged by Sargon early in the eighth century. We thus see that all the expeditions of Jews and Phœnicians were confined to a period of little over a hundred years (1020–900), whereas those of the Himyarites may have extended over several millenniums during the flourishing epochs of the Minaean and Sabaean empires. At any rate we have one date, 1100 B.C., astronomically determined (see above), and how long the traffic in gold and the other costly merchandise may have been established before that date, it is impossible to say. But it came to a close with the Axumite conquest of Yemen about 530 A.D., followed, a century later, by the Moslem irruption, when the Yoktanides went down before the Ishmaelites, and the Himyarites disappeared from history.

They were succeeded by the Muhammadan Arabs, whose flourishing settlements—Magdosho, Mombasa, Melinda, Quiloa, &c.—were found (and mostly destroyed) by the Portuguese pioneers extending along the eastern seaboard, all the way from Cape Guardafui to Sofala. We thus see that the South Arabians were dominant in the Indian Ocean throughout historic times, and that they were the chief forwarders of the sea-borne traffic until they were ousted by the western peoples—Portuguese, Dutch, and English. 'Furthermore, we know that the Red Sea was bristling with activity centuries prior to our epoch; that Arab ships brought spices from India, the cassia tree from China; and the wealth of Arabia in those days was proverbial. The Bible is full of allusions to it, the monuments of Egypt bear equal testimony to the wealth of the people of Punt. Aristeas tells us that a large quantity of spices, precious stones, and gold was brought to Rome—*διὰ τῶν Ἀράβων*, not from Arabia, but "by the Arabians." They were, in fact, the carriers of the ancient world' (Bent, 'Geographical Journal,' 1893, p. 132).

So also Sir Thomas Holdich, than whom it would be

impossible to appeal to a higher authority: 'Although I have come across no traces of Arab occupation on the coasts of Persia, or in Makran, that are *clearly* pre-Mohammedan, it is historically certain that the Arabs were present in large numbers, both in Sistan and Makran (probably also in India) long before their invasion of Sind, early in the eighth century A.D. The Arabs had long been navigators of the eastern seas before that time; nor is there any reason to suppose that the Arabs were not as early navigators as the Phœnicians or the Greeks. By the time that Vasco da Gama rounded the Cape, the Arabs were able to take latitude observations by using the astrolabe, and could get their longitude from observations of the moon's eclipse, and they had learnt the use of the compass from the Chinese, whilst they apparently [read *certainly*] borrowed their system of notation from India. That the Arabs were the first astronomers and the earliest navigators is, I think, also supported by the fact that all the early ocean-going ships of Europe were built on the lines of the Arab "buggalow;" many of our naval terms are Arabic (such as admiral, barge, dinghy, and even jolly-boat), and the names of the constellations are also Arabic. If the Arabs were not the first navigators, they obviously invented their own system of navigation, and borrowed it from no one. The Aryans never were sailors. The Dravidians certainly were, but they went eastward to Burmah and the Straits, and not to the west. The position of the coast Arab city of Tizin in Mekran, and of others on the south coast of Arabia, almost proves the existence of a very ancient trade along the east coast of Africa, a theory which is strongly supported by the evidence of the Periplus' ('Geograph. Jour.,' April 1899, p. 395).

Hence the Arabs must at all times have been well acquainted with the Rhodesian goldfields, and they were able to describe the lie of the land and its running waters to the early Portuguese adventurers in such a way that De Barros calls the 'Kingdom of Sofala' an island, 'enclosed like an island by two arms of a river' (*supra*, p. 3). He took this expression probably from the Arabs themselves, whose *Jezîrat*, 'island,' had a very elastic meaning, being commonly applied by them to peninsulas, as well as to any inland tracts enclosed by converging streams.[79] Thus Senaar between the White and Blue Niles, and Mesopotamia between the Tigris and Euphrates, are 'islands,' as was also Rhodesia, the auriferous kingdom of Sofala, roughly enclosed between the Zambesi and the Limpopo. The early Himyaritic traders must therefore have reported to their Israelitish neighbours that the gold brought by them to their emporium of Ophir came from an island, a region enclosed by ramifying watercourses, and that the gold was of fine quality, as it still is, the proportion averaging in some districts as high as two or three oz. to the ton, and that besides gold there were also precious stones—bdellium and the onyx. All this corresponds exactly with Gen. ii. 11, 12, the literal version of which is: 'Name of the first [river] Pîshôn—it [is] the surrounder of all the land of the Hawîlāh where gold [is]. And gold of that land [is] good—there [is] the bĕdolaḥ and stone of the Shosham.'

We have made the circuit of the Indian Ocean from Rhodesia round by Punt and Arabia to India and the Golden Chersonese, and so back again to Rhodesia, and have nowhere found any land except this which at all corresponds with the Biblical description. Although in its present form Genesis is a comparatively late document, it none the less perpetuates the oral traditions of the Israelites, who had only a second-hand knowledge of the auriferous Havilah itself, hence confusedly associated it, as we have seen, with the gold distributor Ophir, but at the same time accurately described its salient geographical feature, its insular formation, from the reports of its Ḥimyaritiç colonists—builders of its monuments and workers of its rich goldfields. Havilah and Rhodesia are therefore one, and if any still doubt, the burden will fall upon them of producing any other region which will even remotely answer to the known conditions of the problem.

But they will have to do much more than this. They will have to explain away the 115,000 square miles of Rhodesian monuments and ancient gold workings themselves, which on their assumption of some other Havilah would be left out in the cold. But that is impossible. The monuments and the workings, with their slave-pits, their Zimbabye Zodiac, their objects of Semitic worship and other accessories, and above all their vast extent and successive architectural periods, are established by three decades of intelligent exploration, and are no longer open to question. Hence what was tentatively put forward as a plausible, a working hypothesis by the pioneer investigators, notably Mauch,

Maund, Bent and Swan, has ceased to be a subject of speculation, and must henceforth be accepted with Schlichter, Hall, Neal, Johnson, and all other living observers, as a permanent acquisition of human knowledge, one of the most solid and brilliant additions that have been made in modern times to the sum of historic lore.

We thus arrive at the following important conclusions, which I trust may now be considered fairly well established, and may therefore legitimately take the place of the many theories and speculations hitherto current regarding the 'Gold of Ophir,' its source and forwarders :

1. Ophir was not the source, but the distributor of the gold and the other costly merchandise brought from abroad to the Courts of David and Solomon.
2. Ophir was the emporium on the south coast of Arabia which has been identified with the Moscha or Portus Nobilis of the Greek and Roman geographers.
3. Havilah was the auriferous land whence came the 'gold of Ophir,' and Havilah is here identified with Rhodesia, the mineralised region between the Lower Zambesi and the Limpopo—Mashona, Matabili, and Manica lands.
4. The ancient gold workings of this region were first opened and the associated monuments erected by the South Arabian Himyarites, who were followed, in the time of Solomon, by the

Jews and Phœnicians, and these very much later by the Moslem Arabs and Christian Portuguese.

5. Tharshish was the outlet for the precious metals and precious stones of Havilah, and stood probably near the site of the present Sofala.
6. The Himyaritic and Phœnician treasure-seekers reached Havilah through Madagascar, where they had settlements and maintained protracted commercial and social intercourse with the Malagasy natives. With them were associated the Jews, by whom the fleets of Hiram and Solomon were partly manned.
7. The Queen of Sheba came by the land route, and not from over the seas, to the Court of Solomon. Her kingdom was Yemen, the Arabia Felix of the ancients, the capital of which was Maraiaba Bahramalakum. Her treasures were partly imported—the precious metals and precious stones—from Havilah and its port of Tharshish to Ophir, and partly—frankincense and myrrh—shipped at Ophir from the neighbouring district of Mount Sephar.
8. Sephar was confused by the Alexandrian authors of the Septuagint with Ophir, which was the chief emporium of the Sabaean empire.
9. In a word the 'Gold of Ophir' came from Havilah (Rhodesia), and was worked and brought thence first by the Himyarites (Minaeans and Sabaeans), later by the Jews and Phœnicians, the chief ports engaged in the traffic being Ezion-geber

in the Red Sea, Tharshish in Havilah, and, midway between these two, Ophir in South Arabia.

10. This central position of Ophir explains how it became the intermediate emporium whither the fleets of Hiram and Solomon sailed every three years from Ezion-geber for the gold imported from Havilah, and for the spices grown on the slopes of the neighbouring Mount Sephar, not far from the deep inlet of Moscha, round which are thickly strewn the ruins of Ophir.

11. These and the other Himyaritic ruins of Yemen show striking analogies with those of Rhodesia, while the numerous objects of Semitic worship, and the fragments of the Himyaritic script found at Zimbabye and elsewhere south of the Zambesi, leave no reasonable doubt that the old gold workings and associated monuments of this region are to be ascribed to the ancient Himyarites of South Arabia and their Jewish and Phœnician successors.

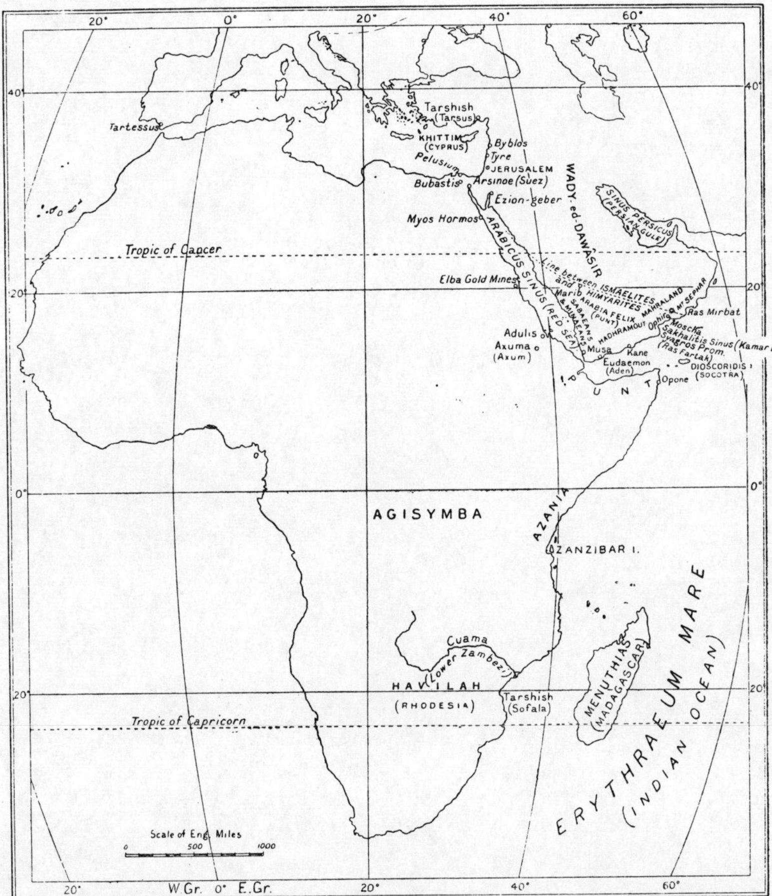

Stanfords Geogr. Estabt. London.

NOTES AND ADDENDA

CHAPTER I

NOTE 1, p. 3.

De Barros' spelling is right, *r* and *s* interchanging in the South African Bantu dialects, as in Sechuana *Morimo*=Zulu *Mosimo*. In Zulu-Xosa there is no *r* sound, which is always replaced by *s* or *l*. The *r* that occurs in Zulu texts is a click. These letters interchange in the Aryan languages also, but always from *s* to *r*, never from *r* to *s*. Thus Lat. *genus*, *generis* for *genusis*; Ger. *Bücher* for *Buches*; Eng. thou *art* for *ast* (*cf.* Sans. *asi*).

NOTE 2, p. 4.

That is, the Ba-Tongas or Ama-Tongas, now in Ama-Tongaland, between Delagoa Bay and Zululand. There are still a Ba-Toka people east of the Ruenya river, who were widespread before the arrival of the Mashonas, and gave their name to De Barros' Matuca Land, *i.e.* the present Mashonaland. It is obvious from the tribal names that in De Barros' time all the Rhodesian populations were Bantus. But see p. 28.

NOTE 3, p. 4.

Massi Kesse, the present chief station in Manicaland, is about 170 miles or 50 leagues due west of Sofala.

Note 4, p. 4.

That is, 'Infidels' (Arab. كافر Káfir, pl. كفرا Kufra), the name everywhere applied by the Arabs to their pagan neighbours; hence the Asiatic Kafiristan (Kafirland). In South Africa this word has been adopted from the Portuguese by the Dutch, and from the Dutch by the English, and has thus become the collective name of the natives everywhere south of the Zambesi. The result is incongruous in the case of many Bechuanas, Basutos, and Ama-Xosas, who are thus still called 'pagans,' although really earnest Christians. Nor is the epithet any longer applicable to the Kufra Oasis of the Sahara, which was so named because when first reached by the Arabs in the eighteenth century its Tibbu inhabitants were heathens, whereas all are now good Muhammadans.

Note 5, p. 6.

For an explanation of these terms I must refer the reader to my monograph on 'The Portuguese in South Africa,' in R. W. Murray's 'South Africa from Arab Domination to British Rule,' Stanford, 1891, p. 13.

Note 6, p. 7.

This identification, based on the likeness between *Agy-Symba* and *Symba-oe*, cannot stand, because Ptolemy's Agizymba was not a fortress, or a palace or a city, but a vast region (*regio Aethiopum latissime extensa*), which he places south of the *Lunae Montes* ('Mountains of the Moon') about the equator. What Ptolemy's *-zymba* may mean I do not know; but De Barros' *Symbaoe* is rightly explained by De Barros himself as a royal court. In the still current local Bantu dialects the components are *nzimba*, a house, and *mbuie*, a lord; hence *nzimba-mbuie* (zimbabwe) clearly means a chief's dwelling. In the widespread Chinyanja language of Nyasaland *mbuie* still means a prince, and *nyumba*, a house, where *ny=nj*, *nz*, &c., as in *Nyasa*, *Nyanja*, *Nyanza*, &c. See A. Riddel's 'Grammar of the Chinyanja Language,' London, 1880.

CHAPTER II

NOTE 7, p. 11.

'It was really Renders who first discovered these ruins, three years before Mauch saw them, though Mauch and Baines first published them to the world, and they only described what the old Portuguese writers talked of hundreds of years ago.'—E. A. Maund, 'Geograph. Proc.,' February 1891, p. 105. Mauch's account appeared in 'Petermann's Mittheilungen' for April 1874; and Baines' in 'The Gold Regions of South-Eastern Africa,' London, 1877.

NOTE 8, p. 16.

Quoted in my 'Boer States, Land and People' (Methuen, 1900, p. 43), where further details will be found.

NOTE 9, p. 24.

'History,' Book II. I understand there is a copy of the rare anonymous English translation of this valuable work in the Grahamstown Library, where it may be consulted by our South African friends interested in the early history of their adopted land.

NOTE 10, p. 25.

'Report on the Farm,' Lisbon, 1883, quoted in 'Boer States,' p. 44.

NOTE 11, p. 27.

'Ruined Temples in Mashonaland,' in 'Jour. Anthrop. Inst.,' 1897, p. 11. The whole passage is worth quoting:— 'The difference in form of the temples in this part of Mashonaland, and of the southern temples near the Limpopo river, is worthy of notice. It is evident from eight of the temples which I examined that the prevailing southern type is a single wall built on one or more curves, but not forming a complete enclosure; whereas the northern temples are generally if not always complete enclosures, formed often of one curve. This would seem to show a slight difference in the cult, and the difference in cult may have indicated a difference in the principal industry of the people. Certainly there is a wide difference in the soil and geological formation of the two districts, for here the country is fairly fertile, and is auriferous, but the country between the Makloutsi and Tuli is certainly not fertile and has not yet been proved to be auriferous. If there was ever any doubt, certainly recent discoveries at Zimbabwe seem to prove that its builders came there for gold; and yet we find temples of the same style in a country where there seems to be no gold. The only suggestion I can offer is that *they went to the latter country for precious stones.* The country there is generally composed of freestone, traversed by trap dykes, and in the altered freestone occur crystallised quartz minerals of many kinds, and sometimes these are of considerable beauty. Many of the precious stones mentioned in the book of Revelation are found there.'

NOTE 12, p. 27.

'Skizze der Geschichte und Geographie Arabiens,' Berlin, 1890, II. pp. 364–5. Here bdellium is regarded as a gum like styrax (storax) and a tree like almuggîm, and then it is stated that in Dhafâr—the Mahra country in South Arabia, where I locate Ophir—gum is still called *amlokh*, the Hebrew *bdolakh*, which Oppert compares with the Assyrian *budilkhati*, 'spear shaft,' while Hommel asks why should not the same tree yield a costly gum and spear shafts, these being amongst the items at one time paid by the Arabians as tribute to the Assyrians. In Greek, *στύραξ* means in mas. a spear, in fem. a tree, in neut. a resin ('lachryma styracis'). Hommel therefore says one styrax is to the other styrax as the Heb. בְּדֹלַח (*bdolakh* the gum) is to the Assyr. *budilkhi* (the spear). And so Glaser is started on a quest of several pages to find out what tree this can be which yields both gum and spear-shafts, and winds up by handing over to botanists the further determination of the 'Bdolakhbaum,' Pliny's 'arbor nigra magnitudine oleæ,' &c. It would be easy to wax merry over these things, but it would be bad taste. We have always to remember that these commentators are men of the highest integrity, thoroughly in earnest, and superb scholars; consequently entitled to all sympathy and admiration in the present, no less than in the earlier sense of the term.

Note 13, p. 28.

In Hottentot, unlike many Bantu and other tongues, there is no *l*, the usual substitute being *r*. Thus the Portuguese *ouro* and *prata* (*gold* and *silver*) become *wolo* and *palata* in the Congoese of the West Coast. For the Hottentot *mari*= gold, not *money*, see J. G. Kroenlein's 'Wortschatz der Khoi-Khoin (Namaqua-Hottentoten),' Berlin, 1889, p. 231. This authority also remarks that '*r* steht im Nama regelmässig für *l*.' This also is common in Aryan. Thus, Lat. *ulmus* Fr. *orme*; and the *English* are *Angrezi-lóg* in India.

Note 14, p. 29.

A. H. Keane, 'Ethnology,' p. 253, where the last clause is quoted from 'Lichtenstein,' I. p. 400. Here also is quoted the following passage from Adelung and Vater's 'Mithridates':

'Für gewisse Gegenden ist dies völlig erweislich, indem Berge und Flüsse des Landes, wo jetzt die Koosa [Ama-Xosa] wohnen, in ihren hottentotischen Namen den sichern Beweis an sich tragen, dass sie einst bleibender Besitz der Hottentoten gewesen sind' (Berlin ed. 1812, iii. p. 290). I have even traced the Hottentots as far north as Lake Victoria Nyanza, and the Bushmen to Lake Tanganyika ('Man Past and Present,' p. 121–2).

CHAPTER III

Note 15, p. 33.

'The Races of the Old Testament,' Religious Tract Society, 1893, p. 51.

Note 16, p. 33.

In the brilliant but highly venturesome *Einleitung* (Introduction) to his 'Nubische Grammatik.' Here the author finds only two stock languages in the whole of Africa, whereas others have discovered at least a score in Sudan alone.

Note 17, p. 34.

That is, 'The Circumnavigation of the Red Sea,' by which the writer means the Indian Ocean, inlets (Red Sea proper, Persian Gulf, &c.) all included. As this valuable document will be again referred to, I may here state that Glaser ascribes it to Basile, an Alexandrian Greek trading in those waters about the middle or second half of the first century (between 56–67 A.D). The report is full of useful information for seafarers, much of it evidently at first hand. It takes the form of a sort of log-book, of course with digressions and much descriptive matter, kept by the trader himself, who, starting from the Myos Hormos on the Red Sea Coast of Egypt, first takes the seaboard to the right to or beyond the latitude of Zanzibar, then to the left round the Arabian peninsula, the Persian Gulf, and Mekrán (Baluchistan), to the west coast of India, with a sort of telescopic glance at the 'Far East.'

Note 18, p. 36.

Passage preserved by Diodorus Siculus (3, 12–14) and quoted by H. F. Tozer in 'A History of Ancient Geography,' Cambridge, 1897, pp. 186–8. There is an Egyptian papyrus in Turin which shows a map of one of these mining stations, and, as it dates from the time of Rameses II., is the oldest chart in the world. It is reproduced in Reclus, x. p. 285 of my English edition.

CHAPTER IV

Note 19, p. 38.

'Alte Geographie,' § 53. One instance of the etymologies will suffice. The *Thebae Oppidum* or *Tabis* is said to be so called from an Arab tribe *Debai*, and both of these terms are supposed to be 'evidently connected with *dahab*, gold'! Now, the Peninsula was invaded and partly reduced more than once by the Persians, under the Achaemenides, for instance, and again in the tenth century A.D., when Abul Kasim, wazir of Adhad ud-Dowla, conquered the whole of Oman, and founded the present Sheraizi, 'Little Shiraz,' named after the Shiraz of Farsistan (965 A.D.). Why, then, might not *Tabis* be derived from the Persian *tab*, fever, heat, &c., in reference to the hot miasmatic districts along many parts of the low-lying *tehama* seaboard. For such a derivation there is this to be said, that the *tab* ('fever') is there unquestionably, whereas the *dahab* ('gold') still awaits discovery. No doubt the fever-stricken coastlands lie more on the west (Red Sea) side of the Peninsula. But we shall see that some philologists allow themselves the liberty of first equating *Abhîra* with *Ophir*, and then translating it bodily from the Indus to the west side of the Persian Gulf. So they at least cannot object to such an equation as *tab*= *Tabis* on the ground of the geographical difficulty.

NOTE 20, p. 39.

'The Land of Midian,' vol. ii. p. 159 *sq*

NOTE 21, p. 41.

The suggestion reminds one of an eminent statesman's 'hands off,' as if the British fleets could sail up the Balkan highlands to drive Austria out of Bosnia and Herzegovina. And we may here recall Burton's remark that 'Midian is not included by Hebrew Holy Writ in auriferous Arabia' (*supra*, p. 37).

CHAPTER V

NOTE 22, p. 44.

The etymology of course is ἄ-πυρος, un-fired, unfused, unsmelted, *ignis expers*, that is, pure, native, gold in nugget form. Certainly a strange popular etymology for cowherds, in whose country there was no gold at all, beyond perhaps a little dust washed down by the Punjab rivers.

NOTE 23, p. 48.

For חוילה (Hawîlah) we are to read חכילה (Hakîlah), 'wie nahezu selbstverständlich,' p. 323.

NOTE 24, p. 49.

'The Races of the Old Testament,' p. 42.

CHAPTER VI

NOTE 25, p. 53.

In the Septuagint *Ophir* occurs altogether nine times, thrice in the usual form Οὐφείρ (Gen. x. 29; 1 Kings xxii. 48; and 1 Chron. i. 23) answering fairly well to the two Hebrew forms אוֹפִיר (Ophîr) and אוֹפִר (Ophĭr); and six times with the intrusive initial Σ and three variants. Thus: Σωφειρά (1 Kings ix. 28); Σουφείρ (1 Kings x. 11; 1 Chron. xxix. 4; and Is. xiii. 12); and Σωφείρ (Job xxii. 24; and Job xxviii. 16). In Ps. xliv. (xlv.) 9, where it ought to occur, it is wanting. Now to see how the Σ crept in we need but compare any of these variants, but especially Σωφειρά, with the 'unto Sephar a mount of the east' (Gen. x. 30), which in the Septuagint is εἰς Σωφήρα, ὄρος Ἀνατολῶν. Here Σωφήρα appears to be an accusative case governed by the prep. εἰς, so that it is almost identical both with the forms in Job and in 1 Kings ix. 28. It might of course be argued that there is a loss of initial S in (*S*)*ophir* itself, and to that view I can have no objection, since it could not affect my case. But there is the difficulty that both forms (*Sephar* and *Ophir*) occur in the same text of Genesis x.

NOTE 26, p. 54.

Dr. Budge, Keeper of Egyptian and Assyrian Antiquities in the British Museum, informs me that this is the proper phonetic rendering of the hieroglyph, of which several variants such as *âb*, *ab*, *eb*, *ebu*, are given. But all are as near or nearer to the Hebrew than the Sanskrit *ibha*. At the same time all these etymologies lose much of their force when it is once understood that Ophir was an Arabian emporium which might have imported such wares either from Africa or India.

NOTE 27, p. 55.

'King Solomon's Golden Ophir,' p. 43. Despite its philological heresies, this is a useful book, containing some valuable data and many shrewd suggestions.

NOTE 28, p. 56.

'Monomotapa (Rhodesia),' Book I. *passim*. Indeed, I think it may now be said that many of these hitherto mysterious remains are undoubtedly of early Phœnician origin. But there are difficulties still needing explanation, such as the occurrence of similar monuments in other parts of the Mediterranean—Malta, Gozzo, Pantellaria, the Balearic Islands, &c.—some of which appear to date from the Neolithic Age, hence cannot be credited to the Phœnicians, whose migration from the Persian Gulf to the Mediterranean cannot have taken place much earlier than about 3,000 B.C.

NOTE 29, p. 64.

Il. 18, 564:—

ἀμφὶ δὲ κυανέην κάπετον, περὶ δ' ἕρκος ἔλασσε κασσιτέρου.

NOTE 30, p. 65.

See especially the 'Verhandlungen der Gesellschaft für Erdkunde,' Berlin, 1896, pp. 2–9.

NOTE 31, p. 66.

'Der Name Ophir hat nicht das geringste mit mahritisch *a'fur* (أعفر) "roth" zu thun, da diese, wie ich in Arabien genau constatirte, mit غ gesprochen wird. Sprenger's Ansicht ist also zu verwerfen' (*op. cit.* p. 377). What Burton really says is that 'Ofir, which in Central Arabia would be pronounced *Áfir* [*á='ain?*] means "red" in Southern Arabia, and the Mahra tribe of Hadramaut [Dhofar] still call Mare Rubrum "Bahr Ofir" [Red Sea]. . . . In classical Arabic 'Ufr (عفر) = "red" certainly cannot be written with an *alif*. The few passages in the Books of Kings and Chronicles all give אפיר [*Ofir*, not *'Ofir* as Burton transliterates], whilst in Job (xxviii. 6) we find עפרות זהב ('Ofiruth Dahab [read Dsahab]) "dust of gold," or gold ore, written in the plural with Oin [*'ain*]' ('The Gold Mines,' p. 262). Thus there is no substantial difference between Burton and Glaser, and these two great Semitic scholars both agree that the Biblical Ophir and the word for *red* belong to two distinct roots. As so much has been built upon the supposed meaning of Ophir, this otherwise valuable quotation from Burton may perhaps be found acceptable.

Nevertheless many still cling to the Indian Ophir, and Mr. E. G. Ravenstein, while admitting that the Rhodesian temples 'pointed to sun-worship and the worship of the linga' [phallus] and that 'such temples are common enough in Southern Arabia,' adds: 'But they are also common to other parts of the world—they are common more especially in India, and I should like to see the claims of India properly considered in this respect' ('Geograph. Jour.' April 1899, p. 394). But the Indian temples here referred to are totally different both in style and construction, as well as far more recent than those of Rhodesia and South Arabia, which in all respects show the closest analogies and even identities, as will be seen further on. The differences between them and the Indian are so great that they could not have been erected by the same people, and India is thus again excluded by the very argument advanced to support her claims.

Then Sir Thomas Holdich asks: 'Why should India be condemned as a country that holds no gold? I have a good many friends who have reason to hope otherwise, and who will consider this a most disturbing assertion. We know that one of the great difficulties with which gold-mining operations in India have to contend is that there are ancient workings in India in the Wynaad [a Taluk of Malabar], in the Kolar district [between the Godaveri and Kistna rivers], Mysore, and in the basin of the Mahanuddy river, and that these workings are driven so deep that it is difficult to drive any further. Out of them an immense amount of gold must at one time or another have been extracted, and I believe myself that even before the days of Alexander gold was largely found *and used in India* [itals. mine]. I believe, in fact, although no statement as to the existence of gold is made by any of Alexander's historians, yet that I am geologically correct in saying every single river in the Punjab holds gold, except perhaps the Beas' (*loc. cit.*, p. 394–95). Now, in the first place, nobody asserts that India 'holds no gold,' but that the quantity is relatively small, so that very little was ever exported, all being needed to meet the great local demand, in fact 'to be used in India,' as Holdich here candidly admits. But in that case none was left for Solomon and Hiram, who must therefore have gone elsewhere for this commodity. Then it will be noticed that all the auriferous districts here specified lie away to the south or extreme east, so that the cost of transporting the output thence to Barygaza, for instance, would be prohibitive. Gold may be washed down by the Punjab rivers, but surely in very small doses, so small that they escaped the notice of Alexander and his keen-eyed Macedonian followers. They reported ivory, of which Vergil heard (*India mittit ebur*), but not gold, which is never mentioned. Nay, more, its existence in those north-western parts is distinctly denied, so that 'no statement as to the existence of gold is made by any of Alexander's historians' is not a full presentment of the case. 'Among his [Alexander's] followers,' writes Schlichter (*ib.* p. 389), 'were many who would have collected news about a gold-belt in India, provided that the latter was actually existing. But no gold country whatever existed there [this is too strong, and should be *was known to the Greeks to exist there*]; on the contrary, all Indian tribes were almost devoid of gold, and Arrianus, our best and most reliable source of information about the campaigns of Alexander the Great, states as emphatically as possible as follows: Alexander and his army have refuted most of the stories in this direction, with the exception of some who have obviously made incorrect statements. It has thus been ascertained that all the Indians through whose territories Alexander and his army marched (and he marched through many of them) HAVE NO GOLD.' Therefore it is evident that 'INDIA CANNOT POSSIBLY BE THE OPHIR OF ANTIQUITY' (p. 390). Schlichter had already remarked (p. 389) that 'at the time when Lassen, Gesenius, Petermann, and others, discussed the Ophir question, it was from want of definite details that theories of a somewhat unscientific character [he means a postulated Indian Ophir] were started.' The illustrious names here mentioned were those of the foremost champions of the Indian theory, and we now see that they receive no support at all from Holdich's 'Ancient Workings in India.' Moreover these workings could never have been carried on by intruding foreigners with slave-labour, for here are no forts or other defensive works in a coæval foreign architectural style, or in any style, and such would be more needed in India with its denser and more cultured populations than in Rhodesia. The alternative of exchange and barter has already been considered.

CHAPTER VII

NOTE 32, p. 70.

'Die Erdkunde,' vol. xiv. p. 372. The passage is of such supreme importance for our purpose that the serious student may be glad to have the original German text: 'Da die Wohnungen der Kinder Joktans sogar geographisch bestimmt sind, zwischen Mesa oder Mescha (Musa) bis Sephar (die älteste Σαπφάρα, die spätere Dhafar, Dhofar bei Mirbat im Weihrauchlande, jetzt Isfor der Eingeborenen, an dem "Berge gegen den Morgen," das heutige hohe Weihrauchgebirge Faguer in der Ehkili Sprache), so ist die Localität der xii Stämme der Joktaniden, zu denen Ophir gehört, ermittelt. Sie trifft merkwürdig genug mit dem Lande Mahra im Osten des eigentlichen Hadramaut, mit dem seit urältester Zeit berühmten Weihrauchlande, zusammen. Darin lag aber Dhofar (Zaphar, auch ein Doffir hörte Niebuhr nennen) die antike Residenz der Himyariten Könige zur Zeit des höchsten Glanzes dieses Reiches (civitas eximia Taphra noch bei Ammian. Marcell. xxiii. 6, 47), welche zugleich seit frühesten Zeiten als EMPORIUM in grossem Waarenverkehr mit dem weiten Oriente stand, und zunächst als jenes Ophir angesprochen werden könnte.'

NOTE 33, p. 70.

The results of Bent's investigations formed the subject of a paper which was read before the meeting of the British Association at Ipswich in 1895, and published in the 'Geographical Journal' for August 1895, under the title of 'Exploration of the Frankincense Country, Southern Arabia.' To these may be added 'Southern Arabia' (1900) a posthumous work edited with addenda by Mrs. Theodore Bent, but containing few fresh data on the subject under consideration.

NOTE 34, p. 73.

'*Op. cit.*' p. 276. The passage referred to in the 'Periplus,' § 30, runs thus: *ὑψηλὰ ὄρη πετρώδη καὶ ἀπόκοπα ἀνθρώπων ἐν σπηλαίοις κατοικούντων*, that is, 'high rocky and secluded mountains occupied by people dwelling in caves.'

CHAPTER VIII

Note 35, p. 76.

Dean Vincent, 'Commerce and Navigation of the Ancients in the Indian Ocean, 1807,' vol. ii. p. 335. This learned commentator, who has many shrewd remarks worth remembering, thinks Ptolemy's *τῶν πέραν ἐμποριών* refers to such Eastern marts as 'Barygaza, Scindi, Oman, and Persis' (*ib.*).

Note 36, p. 77.

'Ruined Cities of Mashonaland,' p. 148.

Note 37, p. 80.

'Animals Mentioned in the Bible' (Religious Tract Society, 1888, p. 174).

Note 38, p. 81.

'Quia classis regis per mare cum classe Hiram *semel per tres annos ibat in Tharsis*, *deferens inde* aurum et argentum, &c.,' that is, '*once every three years went to Tharsis*, *bringing thence gold and silver*, &c.' So also the Authorised Version partly right and partly obscure in 2 Chron. ix. 21: 'For the King's ships *went to Tharshish* . . . every three years once came *the ships of Tharshish* bringing gold, &c.' Here again the Vulgate is right and clear throughout: 'Siquidem naves regis *ibant in Tharsis* . . . semel in annis tribus; *et deferebant inde* aurum, &c.'—went to Tharsis and brought thence gold, as in Kings. It may here be remarked that where Tharshish (Tarshish) occurs in the Authorised Version it is also found, I think, invariably in the Hebrew text, but not necessarily either in the Septuagint or the Vulgate, a fact which has often caused the expounders of Scripture—Cornelius a Lapide, Rosenmüller, &c.—no little trouble. One instance will suffice. In Ezekiel xxvii., verses 12 and 25, the English Version has *Tarshish*, answering to the Hebrew תַּרְשִׁישׁ in both passages. But not so the Septuagint and Vulgate which at verse 12 have Καρχηδόνιοι and *Carthaginenses* ('Carthaginians') and at verse 25 πλοῖα ἔμποροί ('merchant vessels') and *naves maris* ('ships of the sea'). Hence the conjecture that 'ships of Tharshish' had become a commonplace for sea-going vessels, like the famous clippers known as 'Eastindiamen,' which before being superseded by steam were engaged in the China tea trade, but might go anywhere. Nevertheless, we shall see that, although there were several Tharshishes, the Biblical Tharshish of Solomon's time had a very definite meaning.

NOTE 39, p. 82.

Thus Bent: 'The term Moscha, given to it by the "Periplus," is a common term given to bays and inlets on the Arabian Coast. *Merbat*, the name given it by Arabian writers, is still retained in the headland twelve miles east, &c.' The passage occurs in the Report which Bent prepared for the British Association on his explorations in the Dhofar district. Subjoined is the section dealing with the archæology of the frankincense country round about the ruins of Ophir and the adjacent harbour of Moscha: 'The archæological interest in the plain of Dhofar centres chiefly in its connection with the frankincense trade and the towns established in ancient times along the coast by the merchants who provided the ancient world with the odoriferous drugs.

'We have several classical authorities who refer to this district, notably Claudius Ptolemy, the author of the "Periplus of the Red Sea," Pliny, and a few others. From them we can gather certain definite points, that beyond Ras Fartak and the Sachalites Sinus there stretched a fertile coast line known as the Libaniferous coast. The capital of this district was according to Ptolemy called *the oracle of Artemis* (Μαντεῖον Ἀρτέμιδος), and the city next in importance was called *Abyssapolis*, near which was the harbour, the *portus nobilis*, or Moscha of the "Periplus," where the merchants on their way to and from India used to tarry during the violence of the monsoons.

'Along the whole line of the plain of Dhofar there are no less than seven spots where ruins occur, all indicating towns of considerable size; but on close examination of all of them there can be no manner of doubt that at Al Balad and Robat—which are about two miles from one another and connected by a series of ruins—the capital stood. These places are close to the coast, and nearly in the centre of the line of plain, and consist of the remains of many temples, tombs, and public buildings. The acropolis is well marked with the débris of buildings; there is also a tiny little harbour, evidently only for small craft, across which a chain was discovered, the Arabs say, a few years ago. Then there is a moat round the outer edge of this town, in which water is still found, and bulrushes. The columns still standing form an interesting link, which connects these ruins architecturally with the other ruined sites of the Sabaean world; they are square and fluted at each corner, and with step-like capitals. A further development of this is evidently of later origin, when they decorated the capitals with floral and geometric devices. The columns at Axsum in Abyssinia, at Koloë and Adulis on the coast of the Red Sea, and at Mariaba in Yemen, are all of the same character, and indubitably establish the Sabaean origin of the ruins. One column at Robat we found with a capital decorated on four sides, three sides with intricate geometric patterns, and the fourth with the Sabaean letters *ein* and *T* alternately. No other ruins either in size or architecture on the plains of Dhofar can compare with these, and we can safely say that they formed the capital of the district, which Claudius Ptolemy calls the oracle of Artemis (Μαντεῖον Ἀρτέμιδος), and which in later times was known as Mansura, where dwelt, Yakout tells us, the Prince of Dhofar, who had a monopoly over the frankincense trade, and punished the infringement of it with death. In later times the Persians occupied this spot, in the fourteenth and fifteenth centuries of our era. To them we owe the fact of the disturbance of the old Sabaean columns, and the utilising of them to erect mosques, many of which are standing in a fair state of preservation now.

'We tried to find the site of the oracle of which Ptolemy speaks, but could not come to a satisfactory spot until we visited the mountains, and in the Wadi Nehaz, about nine miles from the capital, just at the foot of the mountains, we found a curious natural hole, about 150 feet deep and 50 feet in diameter. Around this there was a wall of Sabaean origin which had a massive gatepost, and in the immediate vicinity were traces of many ruins. From several points I am inclined to believe that this is the site of the oracle mentioned by Ptolemy. In the first place, this hole resembles in character the site chosen for an oracle in the ancient world, bearing a remarkable resemblance to the holes which existed in Cilicia, the oracles of the Corycian and Olbian Zeus, and several other spots in Greece. Secondly, Yakout tells us that the abode of the Adites was half a day's journey from Mansura, the term Adites generally being given to the adherents of the ancient cult; and, thirdly, because there is no other spot on the plain of Dhofar where one can say there is a probability of an oracle existing.

'Yakout further tells us that twenty parasangs from Mansura was the excellent harbour, frequented by the crafts on the way to and from India, and by the merchants in search of frankincense. The "Periplus" also alludes to this harbour, and calls it Moscha, and Ibn Khaldun also speaks of it as Merbat. As we journeyed along the coast we were constantly on the look out for this harbour, and on the second day, after leaving the ruins of the capital, we reached the village of Takha, in the vicinity of which are traces of many ruins scattered about, but inferior in architecture to those at Al Balad. Next morning we were conducted by the natives round a headland, and there saw a long sheet of water stretching inland, but silted up at the mouth by a sand belt, over which the sea flows at high tide. This same sand belt now separates from the shore a rocky island with traces of fortifications on it. There can be no doubt but that this is the harbour, and the island rock guarded the double entrance to it before the invasion of the sand. The harbour is deep, and extends inland about a mile and a half, and there are many ruins around it. Here we have the *portus nobilis* of the "Periplus," the harbour to which the frankincense merchants came, and it is, as Yakout tells us, just twenty parasangs from the capital. The term Moscha, given to it by the "Periplus," is a common term given to bays and inlets on the Arabian coast. Merbat, the name given it by Arabian writers, is still retained in the headland twelve miles east, where Arab dhows find a shelter during the north-east monsoons, but affording no other harbourage, and Ptolemy's name for this place is Abyssapolis, a name which I consider to be derived from the great abyss which I have already described as existing a few miles inland, and which must have been a conspicuous and well-known object to all merchants who frequented this port. Ptolemy, as it will be seen from his name given to the capital, gave a Greek name or equivalent to the places on this coast, and in naming this place he evidently chose the most conspicuous object in its vicinity. Thus we were able to reconstruct on fairly probable lines the geographical features of this frankincense district, and fix the position of its sites.'

CHAPTER IX

NOTE 40, p. 91.

'Sâḳ el Farwain, das zwar nicht ausdrücklich als Bergwerk bezeichnet ist, liegt circa 1 Tag von Dharîjja, &c.'

NOTE 41, p. 92.

'Man Past and Present,' p. 459 *sq.*

NOTE 42, p. 93.

It was the noted city of Tarsus on the river Cydnus, where there is still, I believe, a place of that name.

NOTE 43, p. 93.

'Man Past and Present,' p. 462.

NOTE 44, p. 95.

See my English edition of Reclus, vol. x. p. 365.

NOTE 45, p. 98.

The very word means *Copper*: Greek Κύπρος; Lat. Cyprus, whence Kelt. copar; old German, Chupfar; Anglo-Saxon, copar, &c. The full Latin expression for the metal is *aes cyprium*, 'Cyprus brass,' and as the Phœnicians worked copper in Kittim, and tin in Britain, they had a monopoly of the alloy during a great part of the Bronze Age.

NOTE 46, p. 99.

Bent, 'Southern Arabia,' p. 71–2.

CHAPTER X

NOTE 47, p. 110.

ס=*s* and ש=*sh*: סבא=*Seba*; שבא=*Sheba*.

NOTE 48, p. 112.

But apparently no farther east in the direction of Oman. 'The empire of the Himyarites, which filled Yemen and the Hadhramaut valley with interesting remains, does not appear to have extended its sway so far east [as Oman]; no Sabaean remains have as yet been found in Oman, nor are there any that I have heard of further east than the frankincense country of Dhofar, over six hundred miles west of Maskat' (Bent, 'Southern Arabia,' p. 49). This is important as showing that the eastern seaboard belonged not to the Yoktanides, but to the Ishmaelites, who built no monuments worth mentioning. They were the 'tented warriors,' &c., and another blow is thus struck at Glaser's Havilah and Ophir on the west side of the Persian Gulf. The builders of the Zimbabyes came from the southern Sabaean lands between Dhofar and the Red Sea. The extent of the remains on this seaboard, and no one yet knows how far inland, is surprising. 'All about the Hagarein district are many traces of the olden days when the frankincense trade flourished, and when the town of Doan, which name is still retained in the Wadi Doan, was a great emporium for this trade. Acres and acres of ruins, dating from the centuries immediately before our era, lie stretched along the valley here just showing their heads above the weight of superincumbent sand which has invaded and overwhelmed the past glories of this district' (*ib.* p. 104). And east of Dhofar—nothing! There are no doubt a few remarkable structures in Oman, notably the stupendous flight of steps carried up the steep slope of the Akabal el Hajar in the Jebel Akhdar range. But these are ascribed to the Persians by Colonel S. B. Miles, in the 'Geographical Journal' for November 1901, p. 476.

NOTE 49, p. 113.

It is apparently the *Mareliabata* of Pliny, and certainly the *Mariaba* of Eratosthenes. It has in recent times been visited by Arnaud, Halévy, and Glaser, and the last-mentioned tells us that it lies about four days' journey east of Sanâ, the present capital of Yemen. 'In the earlier Sabaean inscriptions the city is mostly called Maryab, in the later already Marib (مرب),' ii. p. 20.

NOTE 50, p. 113.

Reclus, vol. ix. p. 431 of my English edition.

NOTE 51, p. 114.

This is a common error of long standing. Marib was never called 'Saba,' and this word was always applied to the land and its people, not to its capital.

NOTE 52, p. 114.

R. M. W. Swan, 'On the Orientation and Measurement of the Zimbabwe Ruins,' in 'Ruined Cities,' p. 146.

NOTE 53, p. 115.

'The Fundamental Principles of Old and New World Civilisations,' Cambridge, Mass., 1901. See also my English edition of the 'Aubin Tonalamatl,' a pictorial Mexican Codex in the Paris 'Bibliothèque Nationale,' London, 1901.

NOTE 54, p. 117.

Kremer, 'Akademie der Wissenschaft,' Vienna, 1890, quoted by Bent, 'Ruined Cities,' p. 155.

CHAPTER XI

NOTE 55, p. 119.

'Quia classis regis per mare cum classe Hiram . . . ibat in Tharsis,' &c.

NOTE 56, p. 122.

'Man Past and Present,' p. 495.

NOTE 57, p. 122.

'Human Origins,' p. 94.

NOTE 58, p. 123.

'Mordtmann und Müller, Sabäische Denkmäler,' Vienna, 1883, with eight plates (facsimiles).

NOTE 59, p. 123.

Op. cit., and numerous papers in various German periodicals. Hommel, Halévy, and Sayce should also be mentioned, as sound Semitic scholars who have utilised these 'petrified documents' in diverse ways.

NOTE 60, p. 124.

Sayce, 'Contemporary Review'; two papers: 'Ancient Arabia,' and 'Results of Oriental Archæology;' both excellent, and followed here.

NOTE 61, p. 125.

Op. cit., p. 90.

NOTE 62, p. 126.

Maspero, 'The Dawn of Civilisation,' third edition, p. 616.

CHAPTER XII

NOTE 63, p. 129.

'Notes on the Languages spoken in Madagascar,' 'Jour. Anthrop. Inst.,' vol. xxv. p. 47. This writer appears to credit the 'Periplus' to Arrian, which of course is impossible, but does not otherwise weaken the force of his argument.

NOTE 64, p. 130.

'The present name of Madagascar seems to be due to a mistaken application or extension of the term originally attributed by Marco Polo to the city of Magdoshu (Magadoxo) on the Somali coast. Nevertheless, by a curious coincidence, this appellation bears a tolerably close resemblance to that of *Malagasy*, the collective name of the inhabitants; while the Hova rulers of the island have under foreign influence adopted *Madagascar* as its official designation' (Reclus, vol. xiii. p. 420 of my English edition). The island and its inhabitants were, however, first known to all Portuguese as *Buqua*, from the Malagasy *buki*, 'mongoose,' because some Swahili traders spoke of the people as *Mtu wa Buki-ni*, *i.e.* 'Men of Mongoose-land.' In the same way the Swahili call the Sakalavas of the west coast *Makalalao*, 'cockroaches,' I suppose from the abundance of this pest in those parts, unless it be a popular etymology of the word *Sakalava* itself.

NOTE 65, p. 132.

'The construction of all the Eastern Bantu languages in East Africa may be said to be the same, but take any one of these [and *a fortiori* any other elsewhere in Africa] and compare it with Malagasy, and it will be found that they have but little or nothing in common. The sum total of the influence of the African and Arab [late or Moslem Arab] languages on the western parts [or any other parts] of Madagascar has been to introduce a number of words, of which some few have been adapted and incorporated into Malagasy, as loan words; the others still remain in the lender's hands' (Last, *op. cit.*, p. 53).

Note 66, p. 135.

Here is a really marvellous instance. In Mentawey, a small island off the south-west coast of Sumatra, fifty is *limo-nga-pulu*, lit. 'five-span-ten,' a span of five tens, *nga* being originally=*sa-nga*='one span.' This connecting particle is, of course, lost in the degenerate Malay (*limapuluh*=5 × 10= 50); but it still survives, after thousands of years of dispersion from the Malaysian cradle, in Samoan, thousands of miles away to the east (*e lima-ga-fulu*), where *e* is a remnant of *sa*, one, and *ga* is pronounced *nga* as in Mentawey. So the two compound words are positively identical, although the connecting 'span' has disappeared from most of the intervening Oceanic tongues. In Malagasy it takes the modified form *m*: *dimo-m-fulu*; and *ng* in the Tagalog (Philippines): *lima-ng-pulu*; but elsewhere appears to be lost. This is what naturalists would call a case of discontinuous distribution, such, for instance, as that of the tapir, which is now found only in Malaysia and South America, but was formerly distributed over a great part of both hemispheres. Only the former existence of extinct animal and vegetable species can often be proved by their remains, while there are no linguistic fossils of uncultured languages such as nearly all those of the Oceanic family. Hence all the greater is the wonder that the kinship of its various members scattered over two oceans can be established by such an abundance of cumulative evidence.

Note 67, p. 136.

'On the Relations of the Indo-Chinese and Inter-Oceanic Races and Languages,' in 'Jour. Anthrop. Inst.,' 1880.

NOTE 68, p. 139.

Rev. L. Dahle, 'Antananarivo Annual,' 1876, p. 81. See also his 'Madagascar, Land og Folk,' Part I., Christiania, 1876, *passim*.

NOTE 69, p. 142.

But they occur also in the extreme south-east, and the terms collected by Flacourt over two hundred years ago in the Fort Dauphin district correspond month for month with our table. Thus 1: Alahamady=Alahamali; 2. Adaoro=Azoro; 3. Adizaoza=Alizoza, &c., &c. 'As these Arabic words have their proper meaning amongst the people on the south-east coast, while that is not the case in the interior, I suppose they have been originally introduced there by the Arab immigrants together with some knowledge of astrology, fatalism, and divination. As their descendants and proselytes went up to the interior and began to divulge the more practical part of their doctrine, they soon found that they could not teach, for instance, their *vintana* system without giving the people some idea of lucky and unlucky days. But the computation of those days was to them so closely connected with the names of the constellations of the Zodiac that they found they could not dispense with them. And as nobody knew anything about the stars in the interior, they thought it more easy to keep the names, but to transfer them to the lunar months, since the moon's appearance and course everybody here knew and still knows. So it was, I think, that we here got month-names, which have no meaning as such, and cannot, in fact, have been introduced here [*i.e.* in the inland districts] *at first hand* by the Arabs themselves' (Dahle, *loc. cit.* p. 79). In other words, the ancient Arabs brought the practice of divination, the art of taking horoscopes, and so on, with the corresponding astrological nomenclature, to the coast districts, whence they were gradually spread over the whole island. The process must have been extremely slow, since the Malagasy have for ages been divided into numerous hostile tribes and nations—Hovas, Betsimisarakas, Betsileos, Sakalavas, Antanalas, Antifinerenanas, Antambaboakas, Antisihanakas, Antimahilakas, and many others—who would not readily adopt the innovating practices of their neighbours except under compulsion. But till quite recently there was no compulsion, the various sections of the population having maintained their tribal and national independence till the rise of the dominant Hova people less than two centuries ago. Many were never brought under the Hova rule, and some are scarcely yet subject to the new French administration. The Sabaeo-Babylonian astrology must therefore date back thousands of years to allow time for its spread over the whole island, which is 240,000 square miles in extent, or just about twice the size of the British Isles.

CHAPTER XIII

Note 70, p. 148.

Asad Abu-Karib, or, as Glaser reads the name from the later Sabaean inscriptions, Abûkarib As'ad, is by this authority identified with the renowned As'ad el Kâmil of the South Arabián pre-Moslem traditions (II. p. 510). It is not clear whether he was really a Jew, or only a Judaising Himyarite who flourished during the second part of the fourth century A.D. About Dhu Nowas there is less doubt. He was the last independent ruler of Yemen, having been defeated and killed by the Axumites after a troublous reign of ten years (515–525). He was a great persecutor of the Christians, who had at that time many flourishing communities in South Arabia, and whose appeals to their Axumite co-religionists brought about the Abyssinian invasion and conquest of Yemen. This, as already stated, was followed about a hundred years afterwards, by the Moslem Arab irruption and final subjection of the Joktanides to the Ishmaelites. But long before this time there were Jews in South Arabia, and Strabo tells us that some five hundred who had accompanied the expedition of Aelius Gallus, formed permanent settlements in the country.

NOTE 71, p. 149.

It may, however, be pointed out that Eusebius, a Christian historian of the third and fourth centuries (264–340 A.D.) is no authority at all on these controversial matters. The same remark applies equally to Eupolemos, an obscure writer of unknown or doubtful date whom he quotes in his 'Praeparatio Evangelica' (see the fol. edition of this treatise, Paris, 1628). Certainly it is interesting here to find mention of an island *Urphe*, taken by Gesenius for *Ophir*, which may very well be the true form of the word, since David himself speaks of his 'gold of Ophir' in 1 Chron. xxix. 4. But Chronicles (the *Paralipomena* of the Septuagint) are much more recent than the time of David, and were partly written at a time when this expression had perhaps already become proverbial. *Cf.* 1 Chron. ix. 1, where the Babylonian captivity is mentioned.

NOTE 72, p. 151.

'In cunctis finibus Israel,' says the Vulgate (2 Sam. xxi. 5).

NOTE 73, p. 156.

'The Early Inhabitants of Madagascar,' in the 'Antananarivo Annual,' 1877, pp. 1 *sq.*

NOTE 74, p. 157.

Referring to the publications of the Malagasy Folklore Society, the Rev. J. Richardson stated so far back as 1877 that 'we have a sufficient number of tales in manuscript to enable us to continue publishing for many months to come. To these histories, fables, songs, riddles, &c., &c., there seems no end, while each province and tribe seems to be in possession of some variation of each tale; and each variation gives us some new idea or some additional incident' ('Antan. Annual,' 1877, p. 102).

There were even 'Gothamites' in Madagascar, silly fisher-folk of the Anorohoro tribe, a branch of the Sihanaka nation in the Alaotra valley, who could not speak the Malagasy language correctly, and of course did everything contrariwise, like their British prototypes. Of them it is told that once when cooking eggs they boiled them for hours to make them soft, and then, finding that they only got harder and harder, threw them away as unfit for food. Others having only one slave, who could not paddle the canoe properly, but made it go zigzag against the stream, cut him in two, putting one half at the prow, the other at the stern, and were surprised at the result ('Man Past and Present,' p. 253). There are stories which might almost seem to have inspired Hans Andersen's 'Ugly Duckling,' and others into which snatches of song are introduced just like our nursery rhymes. Such is that of Ifàra, the youngest and prettiest of three sisters, who is hated and driven from her home by the others because of her good looks and wonderful dreams. After many adventures she escapes from the machinations of Itrìmobè, a horrid monster, half man, half beast, who wants to kill and cook her for his supper. Now the disconsolate maiden is seated on a stone weeping sore with grief for her lost home, when a crow comes by, to whom she appeals for aid:

My pretty crow, my pretty crow,
Thy tail I'll smoothe for thee:
To father's well I fain would go,
Oh! come and carry me.

The crow refusing, because he has a grudge against her, then to the kite:

My pretty kite, my pretty kite,
Thy tail I'll smoothe for thee;
To father's well I'd take my flight,
Come now, and carry me.

But the kite also declining, lastly to the reo, who comes along crying reo, reo, reo, as is his wont :

My pretty reo, come near I pray,
Thy tail I'll smoothe for thee :
To father's well I'd take my way,
Please, birdie, carry me.

Thus at last she gets back and is restored to her father's love, while her wicked sisters are unmasked and disowned (Rev. J. Richardson, 'Antan. Annual,' 1877, p. 107 *sq.*). It is the strange European fancy of these simple tales that will strike the observer. Both the pathos and many of the incidents are much the same, and if their inspiration came from the Aryan world, then it will be admitted that the most likely medium of intercommunication were the Phœnician seafarers who manned the fleets of Solomon and Hiram, and had constant intercourse with the maritime peoples of the Mediterranean and Indian Ocean.

CHAPTER XIV

Note 75, p. 165.

To complete this subject I here quote what Sayce says on the antiquity of the Himyaritic script : 'If Dr. Glaser is right, some of the Minaean inscriptions he has collected go back to the age of the Hyksos in Egypt ; at any rate he and Prof. Hommel have made it clear that the kingdom of Saba or Sheba was preceded by that of Ma'in, and that consequently the Minaean or Himyaritic alphabet has an older history than the Phœnician. It would seem to follow that Prof. Hommel is right in holding that what Prof. D. H. Müller has called the Proto-Arabic alphabet is the source both of the Himyaritic and of the Phœnician, and that this fact will have to be taken into consideration in all future enquiries into the origin of the alphabet. It would certainly explain why it is that the Minaean alphabet possesses what the Phœnician alphabet does not—special characters, not derived from any others, for denoting certain sounds which belonged to primitive Semitic phonology but had been lost in the language of Canaan. It will also explain the names of many of the Phœnician letters which have hitherto been a mystery. It is only in the Minaean, not in the Phœnician alphabet, that the forms of the letters *Aleph*, *Beth*, *Pe*, and others resemble the objects after which they have been respectively named : "ox-(head)," "house," and "mouth"' ('Academy,' November 8, 1890).

NOTE 76, p. 169.

Schlichter, 'Geograph. Jour.,' July 1893, p. 49-50. On this important point the writer further remarks : 'Not far from this place [Me'in], at Berahhish, in the lower Jauf district [where Glaser found the inscribed rock of our frontispiece] he [Halévy] discovered the ruins of an ancient temple which, according to an inscription it bears, was dedicated to "*Attar*, and to the gods of Me'in and Ital, to all the gods of countries and nations, to all the gods of the sea and the land, and of the orient and occident, and to the kings of Me'in." This is an ancient ruin, and surely the people who dedicated a temple to all the gods of the sea and land, and to the gods of the different countries and nations, must have had some intercourse by sea and land with foreign countries' (p. 50). Our readers will remember that the Minaeans were a great nation with a long list of thirty-three kings whose names have been recovered from the rock inscriptions, and all of whom reigned before the Sabaeans rose to power (see Chapter X.). We have also seen that the script found in Rhodesia is of the Minaean type, and all the other evidence as it accumulates seems to bear out the assumption that the first settlers in the auriferous lands of south-east Africa were these Minaeans, elder branch of the great Himyaritic family. Some of the structures and workings must therefore be long prior to the time of Solomon, when the Sabaeans were already in the ascendant, and Me'in had been eclipsed by Maraiaba with its Balkis, 'Queen of Sheba.'

NOTE 77, p. 175.

Of course I am aware that the foundations of the true science of astronomy were laid in Babylonia, and doubtless also in South Arabia, at a very early period. 'An accurate system of dating was quite as important to the merchant or money-lender as a code of laws; it is not wonderful, therefore, that every effort was made to regulate the calendar, and invent a system of registering time which should be at once accurate and easy of reference. To Babylonia and its commercial instincts we may thus ascribe the earliest perception of the value of a chronology' (Sayce, 'Genesis,' Temple ed. 1901, p. ix.). But for the bulk of the people this chronology long remained, and in some places (Madagascar) still remains primarily a system of divination, like the Aztec Tonalamatls with their (inter-uterine) year of 20 months of 13 days each = 260 days. See my English ed. of the 'Aubin Tonalamatl,' London, 1901.

NOTE 78, p. 177.

Although this subject of slave labour has already been dealt with, I may here call attention to Bent's remark, in reference to the crushing process in Egypt and Rhodesia, that 'in any gold-producing quarter of Mashonaland, near old shafts and by the side of streams innumerable, crushing stones are still to be seen, used anciently for a like purpose. In several places there are long rows of these crushing stones, sixty or seventy in a row, which would seem to indicate that the slaves employed in this labour worked in rows chained together. Diodorus tells us of the gangs of slaves employed in this arduous labour by the ancient Egyptians, and of the long dark shafts into which they descended. In the Mazoe district we entered several of these ancient shafts, and it is obvious here that, not only for working the mines, but for the construction of the massive buildings, similar gangs of slaves were employed' ('Jour. Anthrop. Inst.,' 1893, p. 131). It is interesting to notice in this connection that, after a suspension of 5,000 years, it is now proposed to resume operations in the old Egyptian gold workings, or at least in the auriferous district of Upper Egypt which Prof. Sayce surveyed in 1901, and found to be richly mineralised.

Note 79, p. 192.

They even call Arabia itself an 'island,' and Hamdâni, the Arab geographer already referred to, gives his work on the peninsula the title of جزيرة العرب 'Jezîrat el-'Arab'=*Island of Arabia.*

INDEX